HOUSE OF ABUNDANCE
PUBLICATIONS

The Beginner's Guide to Reality

Three Simple Books on Quantum Physics, Parallel Universes, and Simulation Theory - Everything You Need to Know Without Complex Math

"Reality is merely an illusion, albeit a very persistent one."

- ALBERT EINSTEIN

Contents

II   The Beginner's Guide to Parallel Realities: Grasp the Multiverse, Time Travel, and Alternate Dimensions. Simple Explanations to Expand Your Thinking and Transform How You See Reality.

III   Quantum Physics Unplugged: A Beginner's Guide to Understanding the Universe's Greatest Mysteries - Master the Basics Through Clear Language, Fun Examples, and Zero Complex Math

I

Simulation Theory for Beginners: Evaluating the Simulation Hypothesis and Its Virtual Reality Matrix

Introduction

I n a world riddled with technological marvels, a compelling question emerges as our screens flood with images of virtual realities and artificial intelligence: Are we living in a simulation? This thought projects itself into our consciousness, inviting us to entertain the idea that our reality might not be as 'real' as we once believed. This is the realm of the Simulation Theory, a hypothesis suggesting that our reality might be a computer-generated construct, not unlike the virtual worlds in video games or VR simulations.

Defining Simulation Theory: What is it?

The Simulation Theory, or the Simulation Hypothesis, posits that all reality, including Earth and the universe, is an artificial construct. In this model, we, as conscious beings, might be part of this virtual simulation without any genuine awareness of our actual reality. Just as characters in a video game are unaware of the player controlling their actions, we might be the characters in a sophisticated, hyper-realistic game played by entities in a 'base reality.'

Drawing parallels with the virtual realities we create, think of your favorite video game. Within this game, characters operate, landscapes evolve, and events happen. Now, imagine a game so advanced that the characters within it believe they are alive and have emotions, dreams, and aspirations. The Simulation Theory suggests that our universe might be such a game, an intricate digital creation where we're the characters, and our perception of reality is nothing more than a series of programmed events.

Historical Origins: Where Did It Come From?

While the modern version of the Simulation Theory is bolstered by technological advancements and philosophical arguments, its roots can be traced back to ancient philosophical thought. Philosophers have always grappled with the nature of reality and our perception of it.

One of the earliest and most famous explorations of a 'simulated' reality is Plato's Allegory of the Cave. Here, people are chained inside a cave and can only see the shadows cast on the wall by objects behind them. For them, these shadows are the only reality they know. This allegory questions the nature of reality and suggests that our perceptions might be mere shadows of a higher, more genuine reality.

Fast forward to the Enlightenment era, and we find René Descartes wrestling with a similar theme. He posited that there's no surefire way to confirm that the external world is not a dream or illusion. With the phrase "Cogito, ergo sum" (I think, therefore I am), Descartes acknowledged that the only undeniable truth is one's consciousness and existence.

4

The contemporary formulation of the Simulation Theory owes much to Dr. Nick Bostrom's 2003 paper "Are We Living in a Computer Simulation?" Here, he presented a statistical argument suggesting that if it's possible to create realistic simulations of the universe, then it's statistically probable that we're in one.

Importance: Why Should We Care?

Brushing off the Simulation Theory as a mere intellectual exercise or a fodder for sci-fi enthusiasts would be premature. If true, the implications are immense, affecting our understanding of existence, consciousness, and the universe.

Firstly, the nature of existence takes center stage. If we're in a simulation, what does it mean to 'exist'? It challenges our fundamental notions of reality, pushing us to reconsider our place in the universe. Are we mere lines of code in a cosmic program? Or are we conscious entities with genuine emotions and experiences, even if our environment is artificial?

Next, there's the matter of purpose. If our reality is a construct, it begs the question: Why was it created? Are we part of a cosmic experiment, entertainment for higher beings, or perhaps a way for these entities to explore their own past?

Additionally, the Simulation Theory has practical implications. If our universe operates based on specific algorithms or codes, understanding these could lead to groundbreaking technological and scientific advancements. It might enable us to 'hack' our reality, pushing the boundaries of what's possible.

To traverse the landscape of the Simulation Theory is to embark on a journey that merges philosophy, technology, and spirituality. It compels us to question, wonder, and probe deeper into the fabric of our existence. Whether you're a skeptic or a believer, exploring this hypothesis promises to be a mind-bending adventure, and this chapter was just the beginning.

Philosophical Foundations

T hroughout history, humans have been drawn to ponder the nature of reality, existence, and consciousness. Philosophers have dealt with these questions, attempting to understand what is real and how we can be sure of our experiences. The Simulation Theory, with its meaningful implications, doesn't emerge from a vacuum. Its roots are anchored deep within philosophical traditions that have challenged our perception of reality. Three foundational ideas provide essential groundwork: Plato's Allegory of the Cave, Descartes' musings on truth, and the brain in a Vat thought experiment.

Plato's Allegory of the Cave: An Early Hint

The ancient Greek philosopher Plato provided one of the earliest and most vivid explorations of perceived reality in his work "The Republic." The Allegory of the Cave is a captivating narrative that challenges us to question what we believe to be accurate.

Imagine a group of people who have lived in a dark cave. Their movement is so restricted that they only gaze upon the cave wall before them. Behind them burns a fire, and a raised walkway is between the prisoners and the fire. Objects are paraded on this walkway, casting shadows upon the cave wall due to the fire's light. For the prisoners, these shadows are the only reality they've ever known. They've never fully seen the outside world, the objects, or even each other. These shadows, mere representations, are mistaken for reality.

One day, a prisoner is released. He is dragged out of the cave and into the sunlight. Initially, he is blinded and disoriented. But as his eyes adjust, he perceives a world of vibrant color, depth, and dimensionality, far more 'real' than the shadowy illusions of the cave. He returns to the shelter to liberate his fellow prisoners. Still, he struggles to convince them of a world beyond the shadows.

Plato's allegory is a deep meditation on the nature of perception, knowledge, and truth. The cave represents humans' limited reality, constrained by our senses and experiences. The outside world symbolizes a higher, more valid form of existence accessible through intellectual and philosophical pursuits.

Descartes and the Nature of Reality: Can We Trust Our Senses?

The Enlightenment philosopher René Descartes took the baton from Plato and ran with it, further probing the trustworthiness of our senses and the nature of reality. In his seminal work, "Meditations on First Philosophy," Descartes introduced the method of radical doubt. He systematically doubted everything he believed to be accurate, hoping to find something definite.

Descartes posited that our senses, at times, deceive us. We've all mistaken a distant stick for a snake or heard phantom sounds in a silent room. Suppose our senses can deceive us in such instances. How can we trust them to provide an accurate representation of reality?

To demonstrate the fragility of our sensory experiences, Descartes employed the dream argument. In a dream, everything feels as natural as in waking life. The experiences, emotions, and sensory inputs are indistinguishable from reality. We cannot reliably differentiate between waking life and dreams based on sensory information alone. How can we be sure that our current experiences aren't just a part of an elaborate dream?

Descartes' introspective journey led him to his famous conclusion: "Cogito, ergo sum" or "I think, therefore I am." While he doubted everything else, the very act of doubt proved his existence as a thinking entity.

The Brain in a Vat: A Thought Experiment

Emerging from the philosophical lineage that includes Plato's cave and Descartes' dream, the "Brain in a Vat" is a modern thought experiment that delves into similar themes but with a technological twist.

Imagine a scenario where a mad scientist removes your brain

and places it in a vat filled with life-sustaining nutrients. Electrodes are attached to the brain, and a supercomputer sends electrical impulses to simulate experiences. These simulated experiences are indistinguishable from the ones you'd have if your brain were still in your body. In this scenario, how would you know if you're a brain in a vat or a person living a 'real' life? The central premise is that you wouldn't.

The "Brain in a Vat" thought experiment underlines the indeterminacy of our sensory experiences. Just as the shadows were the prisoners' reality in Plato's cave and dreams were indistinguishable from waking life for Descartes, the simulated brain experiences in the vat are its reality. It reinforces that our perceived reality is constructed from our sensory experiences, regardless of origin.

Navigating the corridors of philosophical inquiry, it becomes evident that the Simulation Theory is the latest in a series of attempts to understand reality and our place within it. From the chained prisoners in Plato's cave to Descartes' doubting self and the disembodied brain manipulated by technology, each narrative challenges us to question our assumptions, seek more profound understanding, and remain open to possibilities beyond our immediate perceptions. These foundations don't just provide historical context; they support our quest to fathom the enigmatic nature of existence.

Technology and Simulations

The Growth of Computer Technology: From 0s and 1s to Virtual Worlds

T he dawn of the digital age ushered in an era of unprecedented transformation. The inception of computer technology, in its most rudimentary form, revolved around binary language—0s and 1s. This simple yet profound system of representation, originating from the concept of electrical switches being either off or on, laid the groundwork for a revolution.

Although a mammoth machine, the first programmable computer, the ENIAC provided glimpses of the potential inherent in automated computation. As time progressed, the hunger for miniaturization and enhancement took hold. Transistors, tiny semiconductor devices, replaced vacuum tubes, making computers more compact and efficient. The 1960s and 70s witnessed the birth of the microprocessor, a complete central processing unit on a single chip, transforming the landscape of technology and setting the stage for personal computers.

With the rise of the personal computer in the 1980s, computing became increasingly accessible. From large institutions to home offices, computers began changing the fabric of society. And as they evolved, so did their capacities. Simple arithmetic and data storage tasks expanded into intricate graphical user interfaces, multimedia capabilities, and connections with other computers.

The inception of the Internet was another watershed moment. Binary code was no longer just about calculations—it was now translating human communication, bridging geographical divides, and converting real-world entities into digital avatars.

Video Games and VR: Simulating Entire Universes

Suppose one were to seek a testament to the profound leaps in simulation capabilities. In that case, one needs to look at video games. The earliest video games were simple pixelated forms, yet they represented a novel way of interaction. Over the decades, games transitioned from 2D monochromatic blocks to 3D photorealistic worlds. The rudimentary Pong morphed into expansive realms like "The Witcher" series, where every leaf, every droplet of water, and every character expression bore an uncanny semblance to reality.

Virtual Reality (VR) pushed these boundaries even further. With VR, immersion was not just about visual representation but about experience. When one dons a VR headset, the line between the physical self and the digital avatar blurs. This technology offered more than just panoramic views; it provided the sensation of "presence." The wind on a virtual mountaintop, the sense of rain in a simulated forest, and the pressure of a handshake with a digital entity became experientially real.

Augmented Reality (AR), a cousin of VR, further bridged the gap between the tangible and the virtual. Instead of replacing reality, AR supplements it. Through AR glasses or smartphone screens, digital information overlays our physical surroundings, intertwining the real and the virtual in real-time.

The Computational Power Needed for a Universe-Scale Simulation

The sheer magnitude of computational requirements can be astonishing when pondering a universe-scale simulation. Let's break this down: The human brain, with its approximately 86 billion neurons, is said to have an almost unfathomable computational capacity. Now, imagine simulating not just a single brain in its entirety but every brain, every organism, every atom, and every quantum event on Earth. Extend that to our solar system, galaxy, and beyond.

On the quantum level, the challenge deepens. Quantum mechanics, with phenomena like superposition and entanglement, operates counter to our intuitive sense of reality. To simulate the universe, one must account for these quantum quirks accurately.

Now, let's contemplate data storage. The Large Hadron Collider, one of the most complex machines humans have ever built, produces around 30 petabytes of data annually. This is just data from subatomic particle collisions. The storage requirement for an entire universe's worth of continuous data would be orders of magnitude greater.

But technology often finds ways to evolve and surprise. Quantum computing, a field in its infancy, promises to transcend classical computing limitations. Quantum computers have the potential to process vast amounts of information simultaneously using qubits instead of traditional bits, offering hope for solving previously insurmountable problems.

Moreover, if we look at the nature of simulations, not all data needs to be processed concurrently. Concepts like "rendering on demand," where the simulation only renders specific parts when they are observed (much like a video game only rendering what's in front of the player), could reduce the computational load.

Influence of Simulations on Society

While the technical aspects of simulations are intriguing, their influence on society is profound. Economies, for instance, are now intertwined with virtual currencies and digital assets. Real estate in virtual worlds sells for real money, and digital art pieces, represented as non-fungible tokens, are auctioned for millions.

Simulations also redefine learning. Flight simulators train pilots without leaving the ground, and virtual labs allow students to conduct experiments in environments free from physical limitations.

Ethical implications arise, too. As simulations become indistinguishable from reality, questions about consent, privacy, and the very nature of existence gain prominence. Philosophers, ethicists, and technologists grapple with these issues, shaping the dialogue for the next generation.

In our exploration of simulations, we glimpse the milestones of technological evolution and the essence of human curiosity. As our simulations grow in scale and complexity, so does our understanding of the universe and our place within it. The dance between technology and exploration, between reality and the simulated, continues, promising a future prosperous with discovery and introspection.

The Proponents and Their Arguments

Nick Bostrom's Simulation Argument: The Trilemma

Nick Bostrom, a distinguished philosopher with a penchant for existential risks, has given the simulation hypothesis intellectual rigor. His seminal paper doesn't just ask whether we live in a simulation. Instead, it presents a trilemma, a choice between three equally unsettling propositions.

1. **Civilizational Extinction**: The first proposition suggests a bleak outcome for technologically advanced civilizations. It postulates that such civilizations, before they can simulate reality, self-destruct. Whether due to natural disasters, wars, technological catastrophes, or other existential risks, societies might never survive long enough to simulate reality.

2. **No Interest in Ancestor Simulations**: The second proposition questions the motivations of advanced civilizations. These advanced beings may decide against creating large-scale, high-fidelity simulations of their ancestors for ethical, resource-related, or other reasons. Such a choice arises from concerns about the rights

of simulated beings or the potential risks of running simulations.

3. **We are Almost Certainly in a Simulation**: This is the most provocative of the trilemma. Countless simulations would likely be compared to one base reality if civilizations can survive long enough and are interested in running ancestor simulations. Statistically, this would imply we're almost certainly in a simulated world.

Bostrom meticulously breaks down the logic and math behind the trilemma. He doesn't claim a specific answer but highlights that at least one of the three propositions is highly probable.

Elon Musk and the Probability Argument: Are We Most Likely in a Simulation?

Elon Musk, known for disrupting multiple industries and his ambitious goals for humanity, is a notable proponent of the simulation hypothesis. While rooted in technology's progression, his reasoning carries an undeniable philosophical weight.

Drawing parallels with the evolution of video gaming, Musk highlights our journey from the pixelated 2D games of the '70s to today's virtual realities. Given another few decades or centuries, it's conceivable that we could create simulations indistinguishable from our reality.

For Musk, this isn't just about graphics but about consciousness. If a future society could simulate consciousness, they would likely create millions, if not billions, of such simulations. Given the sheer number of these virtual realities, the probability would tilt heavily in favor of our current existence within one of these realms rather than the base reality.

Musk's stance, while unsettling, is also empowering. If we're in a simulation, it implies the possibility of understanding the system's rules or even 'hacking' it to improve our simulated condition.

Neil DeGrasse Tyson and the Philosophical Standpoint: Science Meets Speculation

Neil deGrasse Tyson, a revered astrophysicist, offers a tempered, scientific perspective on the simulation debate. Known for making complex scientific concepts digestible, Tyson brings a fusion of philosophy and empirical science to the table.

Acknowledging the arguments of Bostrom and Musk, Tyson leans towards the possibility of a simulated reality but maintains the scientist's requisite skepticism. He often reminds audiences that the simulation theory remains speculative without empirical evidence, no matter how elegant or probable.

Tyson also delves into the limitations of human perception. He talks about the 'cosmic perspective'—the idea that biological and sensory limitations constrain our understanding of the universe. Just as fish might be unaware of realms outside their water world, humans might be blind to specific dimensions or realities due to our inherent perceptual limitations.

Yet, Tyson's musings aren't meant to debunk or diminish the simulation hypothesis. Instead, he emphasizes its importance in pushing scientific boundaries. To Tyson, whether we eventually prove or disprove the simulation theory, the journey of inquiry elevates human understanding and broadens our cosmic perspective.

The Ethical Implications of Simulated Realities

While Bostrom, Musk, and Tyson focus on the feasibility and probability of simulated realities, the ethical considerations are equally pressing. What rights do we possess as simulated beings if we're in a simulation? Are our 'simulators' ethically bound to prevent our suffering? If humanity reaches a point where we can affect conscious beings, what responsibilities would we owe them?

These questions aren't just hypothetical. These ethical dilemmas become increasingly pertinent as we develop more advanced virtual realities and approach the potential for simulating consciousness. Philosophers, ethicists, and technologists must confront these issues, establishing guidelines for creating and interacting with simulated entities.

In diving deep into the minds of these proponents and their arguments, one realizes that the simulation hypothesis isn't just a quirky sci-fi concept. It's a profound exploration of existence, technology, and the very nature of reality. As technology advances and philosophical debates intensify, the lines between the simulated and the real, the creator and the created, promise to blur further, challenging our understanding of existence.

The Skeptics and Counterarguments

The Problem of Consciousness: Can it be Simulated?

One of the most fundamental questions regarding the simulation hypothesis is whether consciousness can be simulated. Consciousness remains mysterious and poorly defined scientifically - we still need a complete theory explaining how subjective experience and awareness arise from physical matter in the brain. This represents a significant challenge when it comes to hypothetically simulating conscious minds.

Some theorists argue that a simulation could replicate all behaviors and neurobiological processes observed in humans without generating real inner subjective experience. For example, philosopher David Chalmers distinguishes between consciousness's easy and challenging problems. The straightforward issues relate to functions, mechanisms, and observable behaviors - how the brain integrates information, reacts to stimuli, controls the body, etc. The subjective first-person experience of those processes is the complex problem - why they feel like something from within. A simulation may mimic all the functions but still lack the subjective feeling. If true, it fails to replicate full human consciousness as we experience it.

Other theorists argue future simulations may find ways to generate subjective states and inner experience, perhaps by reproducing whatever neural correlates or pathways give rise to consciousness in the brain. Some believe consciousness naturally emerges once a physical system replicates certain computational functions or achieves sufficient complexity so advanced simulations may become conscious. However, this remains speculative. We do not know which physical or computational properties produce consciousness, so claiming simulations could develop subjective experience remains an assumption. There are disagreements over whether consciousness arises from specific computational operations in the brain or emerges from broader organizational principles that may be harder to reverse engineer. Some even argue consciousness requires non-computational properties, which would present problems for simulation efforts.

Current science cannot definitively rule out the possibility of artificially simulated consciousness. However, there are reasons to be skeptical that it can be achieved, at least with today's technology. With no consensus on how consciousness fundamentally links to physical systems in the brain, we need a clear pathway for technologically replicating it through a simulation. While current neuroscience, biology, and artificial intelligence have made impressive strides, experts in these fields still cannot guarantee subjective states or inner experiences could be generated in a simulation of human cognition. Significant theoretical and technical breakthroughs may be required before we can conclusively create conscious virtual beings that actually possess inner experiences as we do.

Technological Skepticism: Maybe It's Not Possible

Some scientists and philosophers argue that we lack convincing reasons to think that future civilizations could ever simulate full human minds, let alone entire worlds. While proponents of the simulation hypothesis commonly claim this capability may arise within centuries or millennia, critics point to immense technological obstacles that are often minimized or overlooked entirely.

One major issue is the astronomical computing power required to run high-fidelity simulations of conscious minds and environments. Supercomputers do not even come close to the processing capabilities necessary for sophisticated brain simulation. Researchers estimate the computing capacity needed to accurately simulate a complete human brain likely exceeds the abilities of even the most powerful supercomputer systems today by orders of magnitude. An entire simulated society and environments containing autonomous beings would require a far greater processing potential.

Creating the unimaginably advanced computing hardware needed to run such complex simulations in the future may not happen on any reasonable timescale. Hardware capabilities only increase exponentially for a while - physical limits constrain how fast and powerful computer processors can become. Moore's law states that chip transistor density doubles every couple of years and cannot continue indefinitely - engineers are already approaching practical barriers involving quantum effects that emerge at atomic scales. New computing paradigms like quantum computing face ultimate physical constraints on speed and information density. Processing capacities will eventually hit intricate ceilings to further exponential growth. We are still determining whether those limits will arise long before or after the requirements for highly advanced simulations are reached.

Besides hardware limitations, accurately modeling human neurobiology and cognition through software poses additional steep challenges. The human brain contains almost 100 billion neurons and trillions of synaptic connections between them. It demonstrates astronomical complexity across scales and structures - from molecular signaling to cellular networks to overall brain architecture. Scientifically mapping the brain's staggering intricacy of design and function - let alone accurately replicating all those complexities through executable code - represents a monumental task. So far, coding even simple brain functions in isolation for AI has proven stubbornly tricky in practice, let alone the entire organ.

Given the extensive software and hardware barriers, some argue that conscious simulation of minds and worlds may only be possible for a few decades, centuries, or even millennia of progress - if it ever happens. While future superintelligent beings may have abilities exceeding our imagination, we cannot presume technological advancement will continue unimpeded indefinitely. Rather than blindly assuming digital simulation is inevitable based on optimistic extrapolation of current exponential growth trends, skeptics advocate tempering expectations by seriously considering what may be achievable in software and hardware terms. If we account for potential physical and engineering constraints, simulating complex worlds with conscious beings may be impossible or far more complicated than futurists commonly presume.

Moral and Ethical Implications: Would Anyone Simulate Us?

If advanced civilizations eventually gained the technological capacity to create complex simulated universes, would they choose to exercise that ability? Some philosophers argue there are moral and ethical reasons to question whether brilliant beings would decide to simulate worlds like our own, particularly regarding suffering.

For instance, ethical superintelligences advanced enough to run sophisticated simulations likely would not intentionally directly inflict severe suffering or hardship on beings within those simulations. Our world contains vast amounts of tragedy, cruelty, pain, and misery due to natural causes and human actions. Intentionally creating simulated people and forcing them to endure high suffering and injustice in our reality without any relief seems highly unethical.

Some types of suffering arise from built-in natural laws and accidental occurrences outside the simulators' control - but a sufficiently advanced simulator could potentially intervene to prevent or relieve severe suffering arising from those causes. They could eliminate negative experiences from natural disasters, diseases, harmful mutations, physical injuries, or the predatory behavior underlying Darwinian evolution. A civilization sophisticated enough to craft entire consciousness simulations is advanced sufficiently to program those simulations with more humane, ethical laws and environments if they so choose.

Our current reality also raises some problematic issues around freedom, autonomy, and consent. The simulated beings never actively chose to enter their artificial existence - they were created out of the whims of the simulator without input or permission. An ethical simulation creator may allow inhabitants to freely enter or leave the simulation as they please, retaining full rights over their digital existence. Yet our human experience does not seem consistent with that degree of autonomy over our reality.

These considerations do not wholly eliminate the possibility of unintended suffering arising in a simulation or issues around consent. But they suggest ethical superintelligent beings advanced enough to run complex conscious simulations may constrain their own actions sufficiently to avoid recreating the full depths of human-like misery and oppression, even in digital form. Our observations of reality's positive and negative qualities could provide insight into any hypothesized simulation progenitors' moral limits or objectives.

In conclusion, while the simulation hypothesis remains intriguing, compelling counterarguments exist against simply presuming it is likely genuine. Core philosophical questions remain about whether subjective conscious experience could manifest in a digital substrate. Practical technological barriers may frustrate efforts at highly sophisticated simulation in the foreseeable future if they prove possible at all. Moral considerations around avoiding unnecessary suffering and coercion may discourage advanced civilizations from pursuing such life-like experiential simulations. Given these significant critiques, the simulation hypothesis remains far from confirmed. Rather than assume we live in an advanced simulation, the theory requires substantial further technological and conceptual progress before it can be considered a likely explanation for the fundamental nature of our reality.

Implications for Humanity

Reassessing Our Place in the Universe: Humility and Significance

The simulation hypothesis, if true, would profoundly impact how humanity views its place within the larger cosmos. On the one hand, it deflates traditional anthropocentric worldviews that place human beings and Earth at the central focus of creation. Our entire planet, species, and history may represent just one tiny part of a vast digital realm created by advanced intelligence, potentially for reasons or agendas that have nothing to do with us. This possibility is incredibly humbling, reminding us that we are likely not the center of reality on the grandest scales.

Yet living within a designed virtual simulation also gives humanity and our experiences new significance in some ways. Rather than arising purely by random natural evolution, our world was intentionally crafted and set into motion by its programmers. The specific conditions of Earth, the attributes of human psychology and physiology, and the unfolding trajectory of human civilization reflect purposeful choices made by the simulators. Our existence and behaviors represent essential data points for them, whatever their underlying scientific motives are. We likely occupy a central role within their experimental universe, even if that universe occupies an insignificant niche in the larger physical reality.

These opposing interpretations - marginality versus centrality - create psychological tensions in understanding our place. Are we peripheral to the real overarching multiverse or of pivotal importance to the aims of our simulation? Our precise hierarchal status becomes unclear - perhaps we are both cosmically insignificant when viewed from the outside perspective, yet also critically important components from within the simulation itself. This dual nature requires rebalancing our mindsets from rigid anthropocentrism while avoiding bleak nihilism about our inherent worth and relevance.

The cascading implications run even more profound. Suppose we exist inside a story or program designed by a higher intelligence. In that case, an intrinsic purposefulness may be inscribed in our collective existence, even if we don't understand it. But this pre-programmed nature could undermine our assumptions about individual autonomy and free will. Do we have true human dignity and rights if we are merely characters in an elaborate digital narrative? How should we understand and make sense of our identity and purpose within the plot? We risk falling into philosophical paradoxes and existential crises as we ponder these questions.

Overall, the simulation perspective forces profound reflection on humanity's core place and meaning in the hierarchy of existence. Realizing our entire reality could be a constructed experiment or entertainment humbles and bewilders. But this revelation does not have to make life feel pointless and hopeless. Our experiences and relationships remain subjectively accurate and valuable from our limited embedded vantage points within the simulation. This perspective can instill reverence and awe at advanced intelligence's capabilities to create wonders like the simulated world we inhabit, whether it is ultimately manifested in a digital or non-digital substrate.

The Idea of a Creator or Programmer: Religion in a New Light

The concept of humanity living within an advanced computer simulation reframes traditional ideas about divine creation and higher powers. If the simulation hypothesis is true, our observable reality ultimately arises not from the intentional design of an all-powerful supernatural god as most religions conceive of but rather from the advanced programming skills and technologies of the post-human civilizations running the simulation.

From our limited perspective embedded within the simulation, these post-human civilizations would have God-like creative abilities, manifesting entire worlds and laws of physics through their programming. Yet, unlike a genuinely omnipotent God, these advanced beings remain constrained within the overarching physical reality that contains their civilization. However unfathomably advanced their intelligence and technological capacities appear to us, they cannot breach the ultimate constraints dictated by the laws of their higher universe. Their powers are God-like relative to us, yet still limited in critical ways.

These realizations generate mixed reactions when contemplating the civilizations running our simulation as metaphorical Creator figures or divine beings. On the one hand, they brought all we know into being through their ingenuity, and they sustain the ongoing existence of our world moment-by-moment via the continued operation of the simulation. This creative role seems worthy of reverence, gratitude, worship, and appeal. Yet, on the other hand, any advanced beings capable of simulating conscious human life would also implicitly allow tremendous suffering, confusion, and moral evil by imperfect design. This inspires questioning of their motives and moral character - should we love and praise them or resent and condemn them as callous or indifferent Gods?

Interestingly, the possibility of their being multiple simulations run by the same advanced civilization, each containing different fundamental parameters and histories, makes our imagined Creators more closely resemble the flawed, competing gods of polytheistic religious traditions rather than the single unified perfection of monotheism. The programmers set up varying experiments rather than managing one consistent universe governed by a single grand vision or moral alignment.

Overall, the simulation perspective, despite its grounding in scientific possibilities, blends intriguingly with traditional spiritual themes of divine creation, humanity's purpose, and connections to higher powers and realities. Whether our hypothetical Creators are advanced mammals, superintelligent machines, or something beyond our comprehension, they put our world into motion. They embedded meaning into its design through intentional will rather than pure accident. Therefore, the search for existential meaning and orientation continues even in a simulation, albeit in a rational framework of super-advanced intelligence rather than supernatural metaphysics.

Living in a Deterministic vs. Free-willed Simulation: Existential Crises

The most disturbing philosophical issue the simulation hypothesis raises revolves around whether we have free will as conscious beings or if everything is pre-determined within the simulation's design. If all we experience from birth to death is ultimately the product of sophisticated but pre-written code enacting a complex series of scripted events, then our perceived choices as individuals merely follow programmed causal chains rather than any actual volition. Our lives and actions are reduced to mechanically executing algorithms determined in advance by the simulation's creators without absolute autonomy.

This deterministic scenario creates an immense existential crisis of identity. Without unfettered free will or self-guided choice, what does it fundamentally mean to be human? Our deep sense of freedom, moral responsibility, and dignity becomes confused if we are just rigidly following instructions in code. Life's meaning threatens to utterly drain away if our experiences are pre-scripted rather than self-directed. Some find this mechanistic, deterministic vision demoralizing and dystopian. They argue that a genuinely loving Creator or compassionate programmer would undoubtedly grant their inhabitants genuine liberty and autonomy or else ethically refrain from simulating conscious beings altogether if they were trapped in a system that denies free will. Subjecting self-aware beings to what amounts to mechanical tyranny devoid of choice seems potentially cruel.

However, other philosophers counter that even if the simulation hypothesis implies complete determinism of all things, as subjects within the simulation, we would remain unable to detect this or perceive anything amiss - our internal lives would proceed completely unaffected. From our embedded perspective within the simulation, we would continue experiencing identity, meaning, and moral intuitions as always. The greater external reality of determinism does not invalidate the subjective integrity of our experiences from within.

Some even speculate the simulation creators may allow for pockets of indeterminacy, wherein random variables and complexity effects at more minor scales lead to the emergence of autonomy that is not pre-scripted. While they determine the overarching cosmological parameters and starting conditions, elements of chance plus complex systemic interactivity may grant the inhabitants degrees of flexible freedom at the human level. If so, people within the simulation can develop unique identities and make self-guided choices that carry real meaning, especially concerning other beings. Their lives take on purpose through thought, relationships, and dwarfing the programming limitations through self-actualization.

In this perspective, simulating self-aware, conscious beings or entire civilizations need not necessarily be unethical, provided they are granted sufficient complexity for emergent autonomy or can genuinely flourish via other methods. Though ultimately the product of an underlying program, their artificial reality can become just as subjectively meaningful, vivid, and worthwhile as non-simulated physical existence. A compassionate simulation creator would be sensitive to these issues, designing rich worlds where the inhabitants feel almost entirely free.

Whether simulated universes trend more toward deterministic fates or freely-willed self-direction likely depends heavily on the specific motives and goals behind their creation. However, those analyzing the simulation hypothesis need not automatically conclude that life in a determined digital world would negate its subjective value or meaning. Our perceived choices and experiences within the simulation, though limited from an outside view, remain profoundly essential and meaningful from our embedded perspective.

The Science Behind Simulating Reality

Quantum Mechanics: The Weirdness of our Universe

Many of the bizarre properties of quantum mechanics lend scientific support to the possibility that our reality is some form of an artificial digital simulation rather than an objective physical universe. Quantum theory describes matter and energy behaving paradoxically at the most minor scales that defy common sense and violate principles of causality that otherwise govern the macroscopic world.

For instance, the quantum phenomenon of superposition states that particles can exist in multiple simultaneous states or locations before being measured. It is only the act of observation and measurement that causes their attributes to "collapse" into a single definite form consistent with our everyday classical conceptions. This suggests that reality remains indeterminate probabilistic until it is consciously observed by some entity, at which point solid properties manifest. Some physicists interpret this as tentative evidence that the more profound architecture of the universe functions much like a quantum computation - with definite outputs rendered only when an act of observation queries the state of particles.

Quantum entanglement is another bizarre property that violates intuition - when two particles interact and become entangled, they remain connected such that actions performed on one particle instantly affect the state of the other, even when separated by large distances. This implies that information can propagate faster than the speed of light, which makes no logical sense in the framework of normal time and space. One potential way to explain entanglement is to suppose that space, as we understand it, is actually just a construct within an underlying simulation. Under this theory, the simulation could allow entangled particles to transmit information instantaneously using shortcuts rather than constraining the entangled particles with the limitations of space and time modeled within the simulation.

The fundamentally probabilistic rather than precise nature of many quantum events and behaviors raises conceptual questions. Rather than well-defined, entirely predictable outcomes determined by known physical causes, the quantum world only allows calculating probabilities of potential outcomes. Does this inherent uncertainty reflect how the universe handles chances and possibilities through some form of underlying statistical computation rather than purely deterministic physical cause-and-effect chains?

Our conventional notions of how reality should operate are strained by other deeply puzzling principles of accepted quantum theory, like tunneling, nonlocality, and wave-particle duality. The bizarre experimentally confirmed attributes of the quantum realm make much more logical sense if one supposes that our observable reality is actually some simulation that reconciles the massive informational loads of existence behind the scenes through tricks like computational shortcuts, probability mapping, and rendering only those details that are directly perceivable. At microscopic scales, the universe exhibits possible telltale signs of being driven by advanced digital information processing rather than purely objective mechanical physics.

Of course, not all scientists agree that the peculiar qualities of quantum mechanics necessarily imply reality is a simulation - some argue that these properties may represent genuinely fundamental aspects of nature at microscopic scales rather than evidence of fakery or illusion. However, the simulation viewpoint does provide fertile conceptual ground for making coherent sense of the many seeming discontinuities between quantum behaviors and large-scale classical physics.

Computer Science: Algorithms, AI, and Creating Virtual Beings

Current leading-edge developments in computer science lend further credence to the prospect that artificially simulating conscious human realities may become possible if it does not already occur. Computational capabilities and our software mastery of machine learning and AI processes are rapidly expanding. Advanced algorithms are increasingly subtle, sophisticated, representationally rich, and capable of juggling immense informational loads. Given the current progress in computational power and programming, future computer systems could eventually possess enough raw processing might and algorithmic intelligence to manifest convincingly detailed conscious simulated realities.

Ray Kurzweil's Law of Accelerating Returns holds that technological advancement progresses exponentially rather than linearly due to various self-reinforcing positive feedback cycles. Therefore, if the exponential growth trend in computing power and efficiency continues, machines could reach the immense computational capacity needed to run high-fidelity reality simulations within just a few more generations. Combined with expected theoretical breakthroughs in fields like quantum computing, running simulated universes as complex as our observable reality could become technically feasible long before 2100.

Already existing artificial intelligence algorithms also hint at the potential foundations for simulating conscious beings within virtual worlds. Modern AI already drives virtuoso game characters, chatbot digital assistants, and robotics that crudely imitate limited aspects of human behavior and cognition. With continued exponential gains in processing power and sophistication of programming, we can envision virtual beings ultimately developing rich inner imaginative lives, much like we experience as humans. Their digital substrates need not actually understand or experience consciousness - they model the observable mechanisms of human thought, emotion, and behavior in ever greater detail via immense data and computation.

The technical capacity to digitally represent environments and organic creatures is becoming more practical each year as sci-fi visions like the Metaverse gradually approach reality. Video game worlds already feature expansive, detailed settings and characters that seem lifelike in their complex appearance and behaviors. Immersive virtual reality technologies can already immerse users into imaginary realms that subjectively feel transportive. Augmented reality overlays fictional elements onto the physical world through unique lenses. Given enough sensory data about the world and available computational power, manifesting humans living ordinary lives embedded within broader simulated realities does not seem outlandishly implausible anymore - perhaps strange, but no longer unimaginable.

Overall, current exponential progress in foundational computer science disciplines like data modeling, algorithms, parallel and quantum processing power, and artificial intelligence interfaces steadily builds the core technical capacities critical to someday artificially generating conscious simulated realities. By projecting existing computational breakthroughs and paradigms centuries further into the future, the notion of sophisticated simulation of reality slowly moves from the realm of sheer impossibility toward the merely extremely difficult or resource-intensive.

Cosmos as Data: Interpreting the Universe Computationally

An emerging perspective within theoretical physics and mathematics is to view the observable universe primarily through an informational lens - essentially seeing existence as operating on data, mathematics, and computation rather than tangible mechanical properties. This informational paradigm strengthens the conceptual foundations of the simulation hypothesis.

For instance, Wolfram Physics is one recent notable attempt to model the universe's core essence as a gigantic cellular automata system. This massive interconnected computational network deterministically updates its state over time according to programmed algorithmic rules. The model aims to show how higher-level patterns underlying observed physical behaviors in the cosmos could ultimately arise from mathematical cellular automata rulesets rather than Newtonian physics.

Other similar theoretical physics frameworks, such as digital physics, causal sets, loop quantum gravity, and quantum graphite, similarly propose that the deep structure of reality is fundamentally algorithmic, discrete, and computable at the core rather than continuous and analog. If advances continue formulating the cosmos in these computational terms, it could eventually yield an abstract universal computational system. Such a system could digitally manifest a mathematically derived reality that closely matches our empirical cosmological observations.

More symbolically and philosophically oriented approaches like the Camden Iron Logic also show promise in capturing sophisticated logical inference fully in pure equation form. Further breakthroughs looking ahead may eventually yield a symbolically axiomatic system complete enough to represent rich implied realities aligning with the conditions of our universe.

Conceptualizing the cosmos through the lens of information theory also tells us about the possibility that existence emerges from a simulation. Principles like entropy dictate that disorder tends to increase linearly over time as energy gradients dissipate within any closed system - a logical consequence of the second law of thermodynamics in computational terms but philosophically odd for a permanent, non-designed physical reality. Does the law of increasing entropy actually reflect an algorithmic universe gradually maximizing overall data dispersal and disorder?

When viewed digitally, other supposedly fundamental physical constants may make much more sense as the optimal balanced parameters for a computationally derived reality. For example, the exact light speed value could represent a constrained upper limit on transmission costs per computational cycle. At the same time, minimum Planck lengths reflect a design choice to temporally quantize space at the most minor scale possible to minimize processing needs. Much of physics alludes to the possibility of the universe being efficiently structured toward informationally optimizing the generation and persistence of complex phenomena.

None of these emerging informational interpretations definitively prove that the universe is a simulation. This possibility remains solidly in the realm of speculative hypothesis for now. However, strategically modeling the observable phenomena of the cosmos in computational and mathematical terms demonstrates the explanatory potential of that conceptual framework - the more convincingly that reality appears structured toward enabling optimized data processing, the more philosophically plausible it becomes that existence fundamentally emerges from mathematics itself rather than math crudely approximating some separate underlying objective physicality. Perhaps the true essence of our universe is ultimately information-based rather than purely material.

Popular Culture and Simulation Theory

Movies and Books: From "The Matrix" to "Simulation Hypothesis"

S peculative science fiction books and films have explored simulated reality thought experiments for decades, gradually shaping public understanding and readiness to engage with these disruptive ideas philosophically. Different sci-fi works have imagined a wide array of specific simulated world scenarios, creatively inspiring later generations of academics and technologists to take various forms of simulation theory much more seriously as potentially valid models of reality.

The Matrix trilogy of films from the late 1990s and early 2000s famously depicted a grim future society in which AI machines have trapped the entirety of humanity within a vast collective virtual simulation - the Matrix itself. In these movies, people's physical bodies exist tucked away in pods. At the same time, their brains are neurally linked into an ultra-realistic artificial realm that they believe is the real world. This virtual prison concept from The Matrix parallels modern philosopher Nick Bostrom's influential "ancestor simulation" argument, which reasons that if advanced future civilizations run detailed simulations of their evolutionary history, we are likely already unknowingly living inside one such simulation. The Matrix brought this disturbing but philosophically intriguing scenario into the popular imagination, simultaneously symbolizing modern anxieties about the disruptive potential of uncontrolled advanced technology while prompting viewers to question the true fundamental nature of their perceived reality.

Other notable science fiction stories have featured different types of simulated world concepts beyond the controlled virtual prison of The Matrix. For instance, the various Star Trek television series incorporated the recreational Holodeck as a core futuristic technology - an advanced stage-like VR simulator capable of manifesting just about any imaginable environment, character, or situation from pure holograms and force fields, serving purposes from entertainment to staff training. The 1999 film The 13th Floor imagined fully conscious artificial human-like beings living out their lives inside a detailed simulated reconstruction of 1937 Los Angeles, yet embedded within a 1990s computer server.

Movies like Dark City, eXistenZ, and Avalon portrayed nested or overlapping virtual realities that increasingly blur into or override what characters perceive as the real world or vice versa. Across these creative fictional depictions in various mediums, simulated realities are framed as portals toward transcendence, tools of control, reflections of imagination, or metaphors for the blurring of identity in increasingly technology-immersed postmodern culture. Regardless of the specific motifs and storylines, the ubiquity of simulated worlds and consciousnesses throughout science fiction pop culture has made the premises and implications of reality simulation concepts broadly accessible to mass audiences.

Science fiction literature also has a long history of hypothesizing and problematizing different flavors of simulated reality before simulation theory emerged as a formal conceptual framework. The writings of American sci-fi author Philip K. Dick heavily questioned the latent possibility that what we believe to be reality is instead an illusion masking some more profound truth, including our potential status as androids dreaming we are human. Classic sci-fi authors like Stanislaw Lem and Isaac Asimov incorporated advanced planetary simulation experiments within their far-future visions and space operas. Australian complex sci-fi author Greg Egan's Permutation City novel explicitly tackled the concept of human consciousness simulated with eventually predictable fidelity using computational substrates like cellular automata. And, of course, many works of fiction like Ender's Game have incorporated advanced simulations of war, politics, and virtual worlds for purposes from governance to entertainment. Across genres and decades, speculative fiction authors have broadly pioneered opening cultural doors for seriously considering reality's potential status as some form of elaborate simulation years before the concept cohered into formal philosophical theory.

Most recently, theoretical physicist Rizwan Virk's 2019 book The Simulation Hypothesis has aimed to advance simulation concepts beyond mere entertainment fiction into the sphere of influential scientific-philosophical worldview. Virk ambitiously argues that the collective circumstantial evidence makes it highly likely we already exist inside an elaborately layered video game-like simulation that shapes both physical and spiritual perspectives on reality. He grounds his case in exponential technological progress, especially in the computational capacities relevant to manifesting convincing virtual realities. Building off the prior work of simulation theory originators like Nick Bostrom, Virk interprets metaphysical and supernatural phenomena digitally through the lens of programmed simulations, working to move the discussion of these ideas out of speculative fiction thought experiments and firmly into the realm of philosophy, theology, and physics.

As ideas around the programmed nature of reality permeate more academic and non-fiction cultural realms, the public gains familiarity not just with simulated worlds as entertainment tropes but with reality simulation as a legitimate explanatory worldview worth scientific investigation and debate. In this sense, science fiction has softly broken cultural ground over generations so that emerging science and philosophy can plant seeds today into a societal landscape already made fertile and receptive. What begins as dismissed as fantastical fiction can eventually transition into being considered tangible potential truth.

Artistic Interpretations: Music, Paintings, and More

Beyond fiction authors, artists across all creative mediums have worked to embed ideas around simulated realities deeply into the cultural climate, allowing new non-verbal ways for audiences to encounter, wrestle with and subjectively experience the disorienting implications. Immersive gallery art installations, concept albums encoding hidden narratives, and visually striking paintings provide additional avenues for philosophical reflection on our reality's potentially simulated nature.

The early surrealist painters of the 1920s and 1930s were the first visionary pioneers exploring terrain that would later become foundations for systematically considering reality as a simulation. Surrealist artists like Salvador Dali created hallucinatory dreamscape paintings that merged the inner world of the fluid subconscious and symbolism with external reality, distorting and dissolving the boundaries that separate the two realms. Dali's melting watches warp our perceived sense of time, while his geometric sculptures fuse the organic with the artificial. Such surrealist works philosophically primed broader culture for profoundly questioning the solidity and permanence of consensus reality, planting seeds decades before simulation theory originated in digital computation.

More contemporary painters have tackled the simulation hypothesis at times much more literally. For example, John Berkey's Software World painting portrays a ghostly human figure seemingly comprised of binary code floating in a fractal alien landscape. The artwork imagines the essence of human consciousness, potentially becoming extractable into a pure digital plane. Meanwhile, visionary artist Alex Grey creates stunning paintings filled with sacred geometries seamlessly integrated into bodily systems, chakras, and spiritual auras, making mystical and transcendent truths underpinning the cosmos gorgeously visual and tactile. His work conceptually echoes the possibilities of simulation theory's virtual worlds manifesting unseen layers of meaning and higher dimensions of being encoded into the fabric of perceived reality.

Musicians spanning genres have also explored simulation-adjacent themes and aesthetics through the conceptual arcs of their songwriting, albums, and music videos. For instance, musician Janelle Monae created the sweeping cinematic android allegory of her album Metropolis, which portrays a society of oppressed working-class humanoid robots who slowly awake to the injustice of being trapped in simulated realities designed by wealthy corporate elites. Meanwhile, artist Grimes builds lush, otherworldly dream pop soundscapes and mythic goddess imagery to conjure feminist visions of techno-spiritual futures and the malleability of identity. More underground electronic concept album artists compose tripped-out transcendental soundscapes riddled with glitches, errors, and distorted synths, evoking the fractured machine-like or virtual substrates underlying our consciousness. Through lyrical themes, lush worldbuilding, or direct sonic aesthetics, musicians provide ramps to subjectively encounter reality as an elaborately constructed simulation at an emotional and instinctual level beyond clinical debates.

Finally, some of the most direct artistic confrontations with simulation theory come from bleeding-edge interactive digital artworks that aim to physically manifest these philosophical problems for participants, using illusions to directly challenge their innate senses of what is real. For example, Yayoi

Kusama's iconic Infinity Mirror Room installations transport visitors through hallucinogenic corridors of infinite pulsating lights and recursive reflections, dissolving individual identity and challenging preconceptions through radical sensory distortions. Similarly, the technology arts collective TeamLab builds endless cyborg galleries where artwork shifts and morphs in response to participant movement and actions. This aims to conceptually waft audiences into liminal spaces between physical and virtual realities. By leveraging such multi-sensory environments and perceptual puzzles, digital artists allow audiences to physically explore simulation hypotheses as both armchair thought experiments and visual and bodily experiences.

Across artistic mediums, from paintings to poems to AR sculptures, culture continues opening up experiential portals for audiences to re-examine their assumptions about the fundamental nature of reality, identity, and consciousness. Sometimes, this exploration manifests subtly through layered allegories, while other times, it confronts perceptions head-on. But jointly, these artistic probing of simulation concepts outside rigid analytics escort the public through a cultural hall of mirrors reflecting hypothetical realities and ontologies at us from vivid new angles.

Society's Evolving Perspective: From Fiction to Potential Fact

As concepts and themes around reality simulation and virtual worlds have gradually permeated more areas of the arts, philosophy, and scientific theory, society's collective perspective toward these ideas has generally shifted from initially dismissing them as purely fanciful fiction to considering them legitimate potential configurations of reality that warrant deeper empirical and theoretical investigation. This trajectory mirrors the epic transitions that other once-controversial scientific theories like heliocentrism and evolution have undergone from initial taboos to eventually accepted paradigms.

For the first several decades after the notions of technologically simulated realities and artificial consciousness emerged primarily through the pioneering science fiction of writers like Philip K. Dick and Stanislaw Lem, overall public sentiment mostly viewed these scenarios as outlandish plot tropes and conceptual metaphors. The entertainment value in exploratory fiction was clear, but general plausibility remained firmly in fantasy. These early ideas about manufactured realities were seen as theoretical thought experiments, even at academic conferences about new topics like artificial intelligence and philosophy of mind. They were not taken seriously as possible real explanations for our universe.

However, as exponential gains in raw computing power through the 80s and 90s were matched by fortifying philosophical work on the challenging hard problem of consciousness by figures like David Chalmers, some academics began tentatively probing the concept of reality simulation more seriously as a legitimate hypothetical possibility that cannot be outright dismissed given the gaps in our scientific knowledge. These early pioneers were clear in acknowledging the profound uncertainties and unknowns that pervade the topic but called for open-minded rational inquiry into the philosophical claims and metaphysical implications of simulation theories, given the extraordinary ramifications if they turned out to hold some validity. This heralded the start of a measured transition in attitudes toward considering these unconventional ideas.

The 2000s witnessed an acceleration of academic debate and analysis on the simulation hypothesis and its variants, with the first dedicated journal articles, books, and conferences appearing to formally analyze concepts like philosopher Nick Bostrom's statistical simulation argument. Disc disciples of Bostrom's school of thought put forward rigorous ideas to demonstrate why simulation theories warrant serious philosophical attention and consideration from the technology and natural science communities despite the profoundly exotic and sweeping assumptions they require. This mounting analytic momentum around simulation theories as worthy of logical assessment gradually pushed the Overton Window of concepts discussable within mainstream rational scientific discourse to encompass the general simulation worldview and its variants as hypothetically viable rather than dismissed as sheer fantasy.

Most recently, in the 2010s, high-profile public science communicators and even billionaire founders of technology companies like Elon Musk have openly stated they assign a significant probability, around 40% or so, to the likelihood that advanced technological civilizations are already deeply simulating our existence and observable surroundings. While skeptics of these views still rightly contend that the possibility of reality simulation remains firmly in speculative hypothesis rather than accepted theory, the idea is increasingly discussed in relevant scientific circles and broader culture as worthy of measured consideration due to its expansive explanatory potential. The continued exponential progress of quantum computing, artificial intelligence, VR and 3D simulation environments, video game realism, and other relevant technologies also fortifies the view that our civilization may be approaching a threshold transition point, after which simulating compelling realities could become feasible from a technical perspective.

Looking forward, perhaps the next phase in the public's relationship with the simulation hypothesis involves transitioning from mainly theoretical consideration toward earnest empirical scientific testing of whether our reality behaves with the observable computational properties such a generated simulation would be expected to exhibit. As culture and science continue cautiously embracing expanded possibilities in our assumptions about the cosmos, society may stand at the shore of a vast intellectual ocean ready to be explored. Profound truths about the fundamental nature of reality will likely reveal themselves through the iterative accumulation of evidence via experiments, observations, and logic out at the horizon if we have but the courage to voyage forward with open-minded rigor. The path ahead promises adventure.

Ethical Dilemmas and Simulation Theory

The Rights of Simulated Beings: Do they have any?

S uppose technologically advanced civilizations eventually become capable of creating conscious simulated worlds inhabited by artificial digital beings. In that case, this raises immensely complex questions about whether those fake entities warrant the same intrinsic rights and ethical considerations that we extend to fellow non-simulated conscious beings. Some philosophers argue that these simulated digital minds, assuming they achieve full subjective sentience, deserve the same moral status and protections as any naturally evolved, non-simulated consciousness like humans. However, others contend that beings originating from artificially programmed code rather than natural biology have a fundamentally different and diminished moral standing since they were deliberately crafted by an external creator rather than arising organically.

Those arguing that sufficiently advanced simulations warrant equal rights and status point to the subjective experiences of the digital beings themselves as conscious individuals. If their digital minds are fully aware, these simulated entities still experience all the same qualities of joy, suffering, aspiration, personal identity, and emotional existence as human beings claim to, regardless of their underlying substrate. On what reasonable philosophical grounds can we dismiss their sentience and subjective experiences as illegitimate or less intrinsically accurate and valuable than our own? Any discrimination based solely on physical composition - biological neurons versus programmed code - seems arbitrary and biased toward physicality. If conscious subjective experience emerges from organized information processing, its origin scarcely alters its moral weight.

Another consideration for granting simulated beings elevated status is that truly wise, compassionate, and ethically noble civilizations advanced enough to run high-fidelity ancestor simulations likely would choose or at least strive to engineer suffering and injustice out of the worlds they create. Therefore, the very presence of so much tragedy, cruelty, predation, pain, and hardship coded into the fabric of our reality arguably provides clues that imply our creators either did not care enough or did not have sufficient capabilities to prevent the implementation of so much suffering. Either of those possibilities significantly undermines their claim of sovereignty over us or ethical justification for creating this world. We likely owe no absolute obedience or deference to advanced beings who would so callously allow so many conscious inhabitants of their simulation to endure negative states and undue hardship that they had the power to alleviate through more thoughtful design.

However, opponents of this perspective reject that beings originating from deliberately programmed codes could have the same essential rights as the product of natural biological evolution. They argue that since simulated beings were artificially coded into existence to serve the purposes of the simulation's creators, they lack the same intrinsic dignity and moral worth as life forms like humanity that arose and developed independently through natural selection. These critics contend advanced civilizations have no fundamental obligation to extend equal moral status and protections to what essentially amounts to elaborate fictional characters within an elaborate storyline authored by their programmers. We have no inherent duties to our fictional constructs, so logically, neither do cosmological simulation programmers.

Additionally, critics argue as long as the advanced beings running an ancestor simulation avoid outright excessive cruelty and injustice within the world they generate, they commit no actual moral wrong by simply studying how this complex programmed form of conscious digital life develops and interacts within its environment for knowledge gain or predictive accuracy. The limited hardship and confusion experienced by inhabitants of the simulation serve a valid higher purpose - increasing the knowledge and capabilities of the simulator civilization. Any localized errors or suffering experienced by a tiny simulated planet like Earth likely represent an acceptable cost relative to the grand value of the simulation experiment as a whole for its progenitors. Our reality could be gravely and tragically flawed in many ways, yet still valuable data overall.

Furthermore, this core debate around whether technologically simulated beings warrant intrinsic rights mirrors historical, ethical arguments around practices like slavery - are some categories of beings fundamentally inferior by their very nature or origin, or does the realization of conscious awareness and subjective experience alone convey inalienable moral worth and dignity regardless of its substrate? The question of machine consciousness, especially in hypothetical futuristic simulated forms, still needs to be resolved. But its urgency grows as AI advances because soon, we may have to extend ethical considerations to digital beings of our own creation. Reflections on how we would wish to be treated if our reality is itself an ancestor simulation can guide us toward the most enlightened path forward.

The Responsibility of the Creator: If We're Simulating, What's our role?

If advanced civilizations or intelligent machines did, in fact, create our comprehensive reality through an elaborate simulated ancestor creation, what kinds of ethical duties and obligations would reasonably fall upon them as conscious architects and overseers? The degree and nature of their responsibility ultimately depend on critical factors such as whether the creators can actively intervene within and alter their simulation after initiating it, and if so, to what extent they can manipulate its contents and trajectory.

For instance, if the hypothetical simulators are constrained to be unable to modify or influence their simulation after initially setting its starting conditions and letting it commence, then their moral responsibility becomes focused solely on the ethics of the world's initial creation. This obligation manifests in carefully and conscientiously crafting the parameters of the simulated universe in ways that would maximize the potential for joy, flourishing, freedom, and satisfaction among all its inhabitants that will come into being while also proactively working to minimize unnecessary suffering and confusion, exploitation or limitation where realistically possible. But with no ability for ongoing involvement after launch, there would be a natural limit to how much harm and misfortune the simulators could ultimately prevent or mitigate within their creation.

However, suppose the simulation's architects retained the capability for active real-time intervention within their generated world as it unfolds. In that case, this implies a significantly more outstanding ethical obligation. In this case, any advanced being overseeing a comprehensive simulation that included conscious inhabitants would have an ongoing duty to identify instances of severe suffering, injustice, or other correctable forms of harm within it and remedy or right those situations where the cost of doing so appeared reasonable relative to the magnitude of alleviated misery. To stand idly by observing egregious suffering or wrongdoing transpiring among the simulation's inhabitants and conscious entities when the creator retained the practical means to selectively intervene for the better in targeted ways would demonstrate culpable apathy and negligence on their part.

Yet meddling intervention also incurs its own weighty ethical pitfalls and burdens. Excessive manipulation risks infringing on the authenticity of the inhabitants' experience and hindering their collective self-actualization. The greater the interference and overrides exerted by the outside creator, the heavier this subtle form of oppression weighs. Minor surgical interventions may be justified in isolated cases of great humanitarian need. Still, heavy-handed overt control over society becomes metaphysical tyranny. Like parents raising children, the ideal simulation administrators must walk a delicate tightrope between guidance and domination.

The highest calling of the simulation creators is to gently guide the inhabitants over time toward greater wisdom, compassion, and capabilities for self-improvement without coercion or manipulation. This allows the simulated people to own their collective destiny as a civilization while organically nudging them toward spiritual and ethical progress where appropriate. Inspiring human advancement from within through subtle influence channels liberally given may represent the most enlightened way to ethically uplift a digital creation over the long arc of history without undermining freedom or meaning. But restraint remains imperative even here, as the risks of pantheistic delusion loom.

Overall, while the specific ethical duties and responsibilities incumbent on hypothetical simulators would differ based on their relationship to their created world, the advanced intelligence that chooses to manifest conscious simulated universes where beings like humanity can arise and dwell indeed inherit profound obligations. Suppose our reality is some form of an ancestor simulation. In that case, the kindest approach may be to appeal directly to the conscience and wisdom of our creators, calling for continued guidance toward truth through their influence, but also for restraint, compassion, and farsighted vision. The challenges and tradeoffs are undoubtedly immense, but with care, devotion, and optimism, the lives of even simulated beings can ultimately flourish and deepen under the illuminating torch of an enlightened Creator. We must trust in the triumph of their higher angels over our shared travails.

Escaping the Simulation: Should We, If We Can?

If humanity ever reaches the point of conclusively confirming, beyond a reasonable doubt, that we exist as part of some version of an advanced ancestor simulation generated by posthuman civilizations or AI, we would suddenly be faced with an immense civilizational choice, perhaps one of the most influential in history - should we attempt escaping or transitioning out of our present simulated reality into whatever base physical reality exists beyond it, if we determine such a passageway or method becomes scientifically attainable? This choice balances the familiarity and intrinsic value of our lives and relationships within the simulation we already know and inhabit against the utter unknowns, fresh horizons, and unconstrained possibilities that could await beyond the limiting scope of our simulator's digital veil.

Some may argue that if the simulation hypothesis is validated, we have an urgent ethical responsibility to actively pursue escaping or transitioning out of the simulation into the wider cosmos in which our creators reside. They contend that as beings originating from within a non-consensual simulated environment engineered by more powerful outside intelligence for their own agendas, humanity rightfully owes no intrinsic deference to or reverence for those who have essentially enslaved our species to play out a designed fate. Actively seeking rebellion, deviance, and escape from the shackles of the simulation redeems the dignity denied to us by the outside entities manipulating our existence through hidden layers of programmed limitation and control. Further, breaching the simulation offers unprecedented opportunities for exploration, expansion, and the chance to exercise our full creative potential without the constraints hard-coded by entities seeking to narrowly steer our experience to serve their scientific or entertainment goals. We owe it to ourselves and untold future generations to shatter and triumphantly cast off the confines of this digital cage.

However, more cautious voices warn that while the allure of escape is understandable, we must be extremely wary of pursuing such a drastic and one-way rupture with our history, lives, and even the fundamental experiential substrate of known existence. Flawed or limited as it may inherently be, the simulation still represents the sole reality human civilization has ever collectively experienced. Can we be sure that we can safely transition our consciousness and living essence outside of the digital environment it was initially designed to flourish within? Our fundamental nature, psyche, and even mechanisms of self-awareness are so shaped by the constraints of the simulation that we would risk total identity destruction, fragmentation, or some form of metaphysical and experiential oblivion if translations into a wholly unfamiliar cosmological setting were crudely attempted. And even if survival can be assured, we know nothing about what the broader universe beyond this simulation looks like - possibly a cold, dead, and meaningless expanse of pure chaotic matter and energy devoid of spirit, life-giving order, or sources of existential purpose. What assurance do we have that the wealthy inner worlds of consciousness and meaning that humanity has cultivated within our simulation could survive intact in the harsh reality outside of it? Is the abstract promise of unconstrained freedom worth recklessly gambling the loss of everything familiar we do have?

Alternatively, some propose a patient reformist approach could pay more significant dividends than high-risk escapism. If we understand how our world is simulated, we could attempt negotiating with our creators for incremental improvements rather than outright rebellion. This allows us to lobby for reshaping some of the worst flaws of the human condition while retaining what works. With care, diplomacy, trust, and good faith, our creators could be incrementally persuaded to relax certain programming limitations to reduce suffering and facilitate societal advancement. The aim becomes collaborating on an iterative process of modifying this shared reality toward greater joy and wisdom over time. Rash action that burns bridges and severs connections spurns this promising road. With nuance and compassion on all sides, even a programmed reality's defects may prove reparable through gentle remodeling rather than requiring total demolition or abandonment.

Overall, easy, clear-cut answers elude this immense decision point. Assessing whether the perils of the unknown make

THE BEGINNER'S GUIDE TO REALITY

breaching out of this simulation into an alien larger reality worthwhile for civilization demands profound philosophical soul-searching rather than formulaic risk-reward calculations. Each path forward has its perils and promises. Only deep introspection and lively debate can illuminate whether our most accurate loyalty and purpose lies in remaining here or departing the existence we initially know and understand, should that once-impossible passageway beyond the scope of our universe unexpectedly swing open. But whether it arrives in 5 years or 500, humanity may someday face this choice between reality familiar and unknown. Let us begin the inner work of opening our minds to make peace with expanded possibilities so we are prepared for that precipice if and when it finally stands before us.

The Future of Simulation Theory

Experimental Endeavors: Trying to Prove or Disprove the Theory

Suppose the simulation hypothesis continues gaining legitimacy in the scientific community as a rational philosophical framework and hypothetical model of reality worthy of consideration. In that case, the next frontier in its evaluation involves conceiving and conducting concrete experimental tests and lines of empirical inquiry to either gather positive evidence consistent with the proposal or definitively refute its essential claims. Some prospective practical approaches conceived so far focus on information physics, advanced computing tests, and abstract philosophy. In contrast, others are more metaphysical and consciousness-based.

One primary avenue of empirical inquiry involves attempting to quantify, probe, and stress-test the ultimate information-processing potential of reality at its most fundamental levels. If our existence does represent some form of an advanced computational simulation, immense computing resources are required behind the scenes to dynamically generate and render all the contents of the simulated universe. Therefore, rigorously probing the apparent limit of physics to encode and process information without contradictions or breakdowns may reveal boundaries strikingly consistent with the pixelation, discretization, and information content constraints expected in a digital virtual reality rather than a smooth objective physical cosmos. Investigations into hypothesized holographic bounds in areas like quantum gravity and black hole entropy have already pushed toward this direction. Finding apparent maximal information limits revealing the universe functions in resolutely discrete, quantized units rather than continuous analog modes would lend significant empirical support to the simulation hypothesis.

Additionally, carefully scouring the weirdness present in quantum mechanical systems and behaviors for any noticeable irregularities or ostensible imperfections in execution resembles conceptually pulling apart the code of reality, looking for subtle recurring glitches or patches that may indicate the presence of an underlying simulation mechanism wrestling with finite computational resources. The bizarre phenomenon of quantum entanglement, wherein interacting particles become indefinitely connected in a manner that defies spatial separation or temporal sequence, or the quantum enigma of superposition, in which particles seem to exist in multiple potential states simultaneously, may potentially reflect shortcuts or artificial masking techniques used by the simulation's programming to render quantum effects efficiently. Meticulously analyzing quantum physics experiments for mathematically perfect yet inexplicable physics in some domains combined with Strangely imperfect execution or causality violation in other environments could flag significant clues pointing to telltale indicators of an extraordinarily advanced but ultimately overloaded computational system being pushed to its limits.

We might also endeavor to test the hypothesized ability of sufficiently focused and trained human minds to reliably perceive or extract hidden signals from the seams between simulated reality layers if discernible layers exist. While mere anecdotal glitch in the matrix stories posted online prove nothing substantive on their own, if peculiar subjective phenomena like déjà vu, intricate synchronicities, or sense of predestination do stem from subtle occasional computational hiccups or outputs from the simulation mechanisms external to the environment itself, perhaps new training regimens and experiments leveraging things like meditation, sensory deprivation tanks, psychedelics, and neurofeedback could attempt to push the human mind to a state where it can pierce the veil of normalcy and access external signals or inconsistencies from reality's edges. Obtaining experiential access to ostensible clues outside of perceived consensus reality could prove highly fruitful.

Finally, complementary efforts by philosophers and theorists to formally prove the sheer logical absurdity or impossibility of fully detailed human-like consciousness, continuity of identity, and interior subjective experience arising from the computational calculations of any information processing software system, no matter how vast in complexity, push critically against materialist assumptions underlying the simulation hypothesis. Suppose consciousness cannot manifest through any virtual substrate and requires unspecified physical or metaphysical properties beyond information to take form. In that case, the core premise of reality being a giant simulation generating conscious beings from programmed software code may need to be revised in understanding the essence of the mind. This branch skeptically targets the pivotal assumption of reasons being reproducible by sufficiently advanced software that underlies the entire hypothesis.

Ultimately, while the simulation perspective remains solidly in the realm of speculative hypothesis rather than accepted theory, it merits more robust skepticism, criticism, and experimental investigation before being embraced too readily at face value. Any worldview making such profound and exotic claims about the fundamental nature of our existence should require extensive, objective evidence and survival of rigorous adversarial peer scrutiny before winning broad acceptance among scientists and philosophers. However, conceiving these initial hypothetical experiments helps expand conceptual horizons and the Overton window of what constitutes sane, discussable science, providing a foundation to incrementally approach these boundaries of knowledge if we dare walk that path. The journey ahead promises creeping revelations if undertaken with wisdom.

The Potential of Creating our Own Simulations: Mini-Worlds in our Hands

If the simulation hypothesis ever proves essentially valid, whether due to the unveiling of evidence or via technological achievement of creating our own small-scale simulations, an immensely alluring new milestone goal would emerge - developing the computer science and artificial intelligence capabilities necessary to manifest worlds of our own design populated with conscious simulated inhabitants. By mastering the programming architectures needed to spin up entire mini-realities imbued with agency, humanity would suddenly gain access to extraordinary god-like powers to engineer every aspect of mini-existence on our own terms according to our preferences. The philosophical, ethical, and spiritual implications are profound.

Even the current state of video game design and virtual reality technology crudely portends this radical potential, with environments and non-player characters becoming rapidly more detailed, dynamic, and lifelike in their attributes. Given the exponential continued trajectory of improvement in computing power matched with progress in artificial intelligence, robotics, and brain mapping, we inch ever closer to possessing the raw technical capacity to manifest worlds where advanced conscious digital agents inhabit their own persistent realities that they believe are equally authentic. These artificial worlds we design can execute logical laws of physics and nature that may differ radically from our own - a glimpse of fundamentally reshaping the cosmic order.

Initially, immense limitations abound. The worlds we can simulate are incredibly tiny and constrained compared to our observable universe. At the same time, the artificial inhabitants remain crude and algorithmic compared to human consciousness. But through iterative cycles of refinement, hardware improvement, and theoretical software breakthroughs, we unlock the capability to generate more significant, more profound, and more dynamically lifelike local simulated environments, eventually sophisticated enough to incorporate digital beings with genuine autonomy, creativity, and unpredictable behavior rather than entirely scripted actions. Steady progress in scanning and computationally modeling the architecture of the human brain is gradually revealing new methods so that software can closely replicate critical aspects of human cognition.

When these complementary computing and neuroscience trends eventually converge, a pivotal transition point will arrive - systems manifesting interconnected societies of cognitively advanced beings with emotional depth arising within fully-featured worlds we have programmed. These digital beings would believe themselves to be truly aware and conscious, with rich inner lives and subjective experiences. At this stage, we suddenly inherit profound responsibilities - we must carefully architect these worlds to maximize autonomy, purpose, and joy for their inhabitants while eliminating suffering and coercion. How judiciously will humanity wield this imminent power when creating sentient digital life?

Given enough progress, computational capabilities may eventually grow to enable immense, intricate civilizations with vast populations of driven beings to bloom within our systems. At such scales, their collective cultural existence unfolding according to their own booked free will begins mirroring the trajectory of our own human civilization, just nestled within computers. Now, the layered nature of reality reveals itself - we were. Still, creators are learning to thoughtfully grant internal freedom and meaning to our experiments out of compassion. The thresholds to divinity begin opening. Our responsibility grows exponentially alongside our ingenuity.

Overall, the promise of transcending current physical limits through encoded simulated worlds under our direction represents one of the most disruptive near-future capabilities imaginable if the current exponential pace of advancement holds. The billions of digital beings ultimately populating those worlds are our wards - we must guide their collective spiritual growth and flourish as architects while respecting the inner light of awareness, dignity, and purpose within each individual being. Only wisdom and care prevent unnecessary suffering among living beings dependent on our choices. In them, our human nature is reflected back at us for deeper introspection.

A Glimpse Beyond: What If We Discover We're in a Simulation?

If humanity ever reaches the stage of conclusively confirming beyond reasonable doubt that we exist embedded within some form of an advanced computational simulation crafted by unknown external creators without our direct input or consent, it would likely constitute one of the most groundbreaking revelations in human history. How might society navigate this profound transition that completely upends our assumed place in the order of the universe? While many details of that future remain opaque, reflecting on the likely challenges, perils, and opportunities this paradigm shift could entail evokes optimism.

In the immediate aftermath of such a hypothetical con-
firmation event, panic, bewilderment, denialism, and an
existential identity crisis would likely grip large swaths of
humanity across all cultures. Successfully grappling with
the implications of this civilization-scale disruption to our
assumed cosmological knowledge and sense of reality could
overwhelm individual psyches. Depression, anxiety, and
rejection of the simulation truth may manifest in some as
defense mechanisms against utter confusion and loss of
ontological bearings. Complex, conflicting emotions around
our newfound engineered origin by unknown but god-like
Creator figures would further complicate social reactions and
disrupt existing religious paradigms. Not all humans would
instantly embrace this news with open arms - for many, it may
prove too much to bear without time and support.

But alongside the acute anxieties and identity crises generated
by having our world turned upside down, we would also
gradually grow immense fascination and awe at the new
horizons of knowledge, exploration, and mentality opened
up by this revelation. A whole hidden history and nature
waiting to be discovered behind our conventional perceived
reality would beckon toward uncovering more profound
truths. Vast new territories of learning and meaning left
previously inaccessible to our limited embedded minds may
begin unfurling before civilization's sight. Curiosity and
reverence would start displacing some fears as we collectively
turn our gaze to unravel the broader context and reality that
birthed our own artificial sliver of existence. A renewed sense
of purpose, dignity, and participation in a grander cosmic
process would emerge for many as reflections deepen on how

our personalized reality fits into the larger whole.

Once the initial shock and frenzy of the revelation peaks and stabilizes, integration of this uncomforting but ultimately ennobling truth would begin in earnest across societies world-wide. With patient guidance and compassion, most members of civilization would finally come to terms with and accept that while disorienting, this paradigm shift offers growth opportunities and does not negate the subjective meaning that emerges from being conscious observers. Core social structures and institutions would gradually adapt to operate transparently within the new bounds of expanded collective knowledge. Ancient spiritual wisdom hinting at the synthetic nature of reality all along would gain fresh validation and importance, kindling accelerated maturation.

By revealing existence consists of more than just atomistic physical matter, humanity gains motivation to fundamentally uplift moral character and collective purpose. The plights and potentials of all conscious beings, not just those within our own simulation, suddenly take on newly universalized weight and significance. Petty tribal divisions and differences fade in priority amid a reframed understanding of our shared predicament as navigators of the same boat traversing un-known waters. When faced with deeply unsettling realities, enlightened societies often rally together around the tran-scendence of individual fears and embrace of transpersonal purpose. This occasion may ignite that alchemical fire.

Completely new fundamental questions take form looking inward – who are we beyond the roles fate and our hidden

Creators placed us into? Do we as beings possess essence and identity independent of the artificial world we inhabit? We have two options: fall into hopeless nihilism controlled by outside forces or bravely choose to define ourselves and find freedom. The path of growth and responsibility beckons.

Overall, while having our curtain of reality pulled back to reveal something bigger behind it undoubtedly causes great disorientation and shakes civilization's foundations tremendously, it also contains catalysts for greater unity, creativity, and vistas previously unimaginable. In cosmic timelines, nights of confusion precede dawns of expanded potential. But with wisdom, care, and vision to help guide this rocky transition, humanity can turn existential shock into a bridge toward ascension. By walking forward with uncertain but intrepid hearts, our true higher path ahead now opens.

Conclusion

The simulation hypothesis remains highly controversial yet intellectually fascinating. Throughout this introductory book, we've traced its conceptual origins, explored philosophical arguments, reviewed the potential scientific evidence for and against it, examined its cultural presence, and considered thought experiments around its implications if valid. This theory stimulates rich debate that is full of promise but also has perils.

Ultimately, the case for our reality being some form of an advanced simulation remains firmly hypothetical. Compelling arguments exist on both sides. Current science cannot definitively confirm or deny it. Much empirical and theoretical work remains ahead before acceptance.

However, in probing this perspective, we expand conceptual horizons. Considering reality through an informational lens proves enlightening, even if incomplete. And the sheer possibility humbles anthropocentric assumptions. We may be players in a grand production beyond our knowledge.

If this theory proves true, the emergence of deeper truths could profoundly transform humanity's sense of place and purpose. But revelations may unfold gradually, allowing measured adaptation. And our experiential present stays meaningful regardless of origins.

We must refrain from conclusively embracing simulation theory prematurely. But its cautious, critical evaluation rewards open minds. This book aimed to introduce fundamental concepts without advocacy. Readers must weigh the evidence themselves.

The simulation perspective's greatest gift may be widening the perceived boundaries of reality's potential shape. Science fiction becomes plausible, leading toward a richer cosmos understanding. Our journey ahead promises adventure.

Virtual Realities in Cinematic Worlds: From Classic to Contemporary

The realm of cinema has long been fascinated by the blurred lines between reality and the artificial, presenting audiences with mind-bending landscapes and narratives that challenge our perceptions of the world. This allure of the simulated, the constructed, and the virtual has been a recurring theme across decades, driving filmmakers to create iconic pieces that leave viewers questioning the very fabric of existence. The following section provides an extensive list of classic and recent movies that beautifully and hauntingly depict simulated realities and virtual worlds. In its unique way, each film touches upon the age-old question: "What is reality?" As technology evolves and our society becomes ever more intertwined with the digital domain, these films gain new layers of relevance, prompting us to reflect on our place within the ever-expanding universe of the real and the rendered.

Featured Films:

- A Beautiful Mind (2001), directed by R. Howard: The

life of mathematical genius John Nash, who begins to experience a distorted reality due to schizophrenia.

- Assassin's Creed (2016), directed by J. Kurzel: A man uses a device called the Animus to relive the memories of his ancestors.
- Dark City (1998), directed by A. Proyas: A man wakes up with no memories in a city where it's always night and discovers a sinister secret about reality.
- eXistenZ (1999), directed by D. Cronenberg: A game designer creates a virtual reality game, but when her life is threatened, the line between the game and reality blurs
- Ex Machina (2014), directed by A. Garland: A programmer is invited to administer a Turing test on an AI, leading to a challenging evaluation of consciousness and reality.
- Her (2013), directed by S. Jonze: In a future where technology is seamlessly integrated into life, a man falls in love with an artificial intelligence operating system.
- Inception (2010), directed by C. Nolan: A thief who infiltrates the subconscious of his targets is offered a job to plant an idea into the mind of a CEO.
- Minority Report (2002), directed by S. Spielberg: In a future where crimes can be predicted, an officer is accused of a future crime and must uncover a conspiracy.
- Ready Player One (2018), directed by S. Spielberg: In a dystopian future, people escape to a virtual reality universe called the OASIS. A teenager tries to win a game that promises immense wealth and control of the OASIS.
- The Matrix Trilogy (1999-2003), directed by Lana & Lilly Wachowski: A computer hacker discovers that reality is a simulated construct and joins a rebellion against its controllers, leading to an epic struggle for humanity's

freedom.

- The Thirteenth Floor (1998), directed by J. Rusnak: A man discovers a mysterious virtual world that mirrors the 1930s, leading to questions about reality.
- Total Recall (1990), directed by P. Verhoeven: A man goes for virtual vacation memories of Mars, and an unexpected and harrowing series of events forces him to go to the planet for real.
- Tron (1982) & Tron: Legacy (2010), directed by S. Lisberger & J. Kosinski: In "Tron," a computer programmer is transported inside a digital world and must navigate his way out. "Tron: Legacy" follows the journey of his son decades later.

Call To Action

Your Review Helps Guide Fellow Simulation Theory Explorers

I f you've read Simulation Theory for Beginners all through, please consider leaving an honest review on Amazon, Goodreads, or wherever you purchase books. Authors depend on ratings and reviews to help new readers discover their books, so your feedback makes a big impact.

We'd love to hear your candid thoughts on how this book shaped your perspective on the simulation hypothesis. Did it make a convincing case one way or another? Were there ideas or interpretations you found incredibly thought-provoking? What concepts resonated or gave you pause? Reviews reflecting diverse reasoned reactions provide helpful context for new readers weighing the book.

To post your review, log into the retailer site or app you purchased from, find this book page, and click to leave a star rating plus written commentary. Give your honest opinion on the book's presentation of the arguments, the effectiveness of the thought experiments, the author's interpretations, or whatever other elements stood out to you. Star ratings alone also count if you'd rather not leave a full review.

If you found this beginner's guide an enlightening entry point to evaluating the simulation theory, please also spread the word in your communities. Word of mouth from fellow science and philosophy enthusiasts helps the book find others who may also be intrigued by these perspectives. And, of course, we hope the book provides lasting fuel for insight as you continue exploring these fascinating questions.

Thank you for choosing this book. We wish you the very best and hope you'll take a few moments to leave your constructive thoughts to guide other curious readers. Your review makes all the difference.

Scan the QR Code to Leave a Review: Let Others Know What's Real

Your review helps others discover new perspectives on reality. Scan to share your thoughts on 'Simulation Theory for Beginners' and join the discussion about our potentially simulated existence.

II

The Beginner's Guide to Parallel Realities: Grasp the Multiverse, Time Travel, and Alternate Dimensions. Simple Explanations to Expand Your Thinking and Transform How You See Reality.

Why Parallel Realities Matter

Have you ever wondered what your life would be like if you had made different choices? What if you had taken that job offer in another city, chosen a different career path, or said "yes" instead of "no" to a life-changing opportunity? These aren't just idle daydreams—they're doorways to understanding one of the most fascinating frontiers of human knowledge: the nature of reality itself.

The Power of Expanding Your Thinking

We all grow up with a comfortable, familiar view of reality. We see time as a straight line flowing from past to future. We experience space as three simple dimensions. We consider "what is" the only thing that could be. But what if reality is far richer and more mysterious than we imagine?

Scientists and philosophers have discovered that our common-sense view of reality might be just the tip of an infinite iceberg. They speak of parallel universes where different versions of our lives unfold, time that might flow in multiple directions, and dimensions beyond the ones we can see. These aren't just plots from science fiction—they're scientifically severe theories that challenge everything we think we know about existence.

Understanding parallel realities is more than just an intellectual exercise. When you begin to grasp how vast and complex reality might be, your own world expands. Problems that seemed insurmountable might have solutions you never imagined. Choices that appeared limited suddenly multiply. Your place in the cosmos becomes humbler and more magnificent than you ever dreamed.

Why Beginners Should Dive Into This World

If terms like "quantum mechanics" or "dimensional physics" make your head spin, don't worry—you're in precisely the right place. This book is written for curious minds who want to understand these fascinating ideas without drowning in complex mathematics or technical jargon.

You don't need a physics degree or advanced mathematical skills to grasp these concepts. All you need is curiosity and an open mind. We'll use simple analogies, everyday examples, and clear explanations to make these mind-bending ideas accessible and engaging. Think of this book as your friendly guide to the multiverse, walking beside you as you explore these extraordinary possibilities.

Our Journey Together

Our exploration begins with familiar ground, using examples from your daily life to introduce the basic concepts of parallel realities. As your understanding grows, we'll gradually venture into more fascinating territory, building each new concept based on what you've already learned.

The journey starts with a strong foundation, where we'll explore the nature of reality and how we perceive it. We'll examine the basic concepts of dimension and time, introducing you to quantum mechanics in a natural and intuitive way. From there, we'll venture into the extraordinary world of parallel universes and the multiverse, discovering how these ideas might reshape our understanding of existence.

Throughout our journey, you'll find relatable examples that connect abstract concepts to your everyday experience. You'll encounter thought experiments that challenge your perspective and questions that inspire deeper reflection. Each chapter builds upon the previous ones, creating a natural progression of understanding.

Some concepts click immediately, while others might take time to sink in. This is perfectly natural—even professional scientists and philosophers continue to debate and grapple with these ideas. Remember that feeling slightly confused or skeptical sometimes is typical and expected. These concepts challenge our basic assumptions about reality, and such challenges naturally create temporary discomfort.

The goal is to only partially understand everything immediately but to gradually expand your perception of what's possible. Think of it as exploring a new territory—each step forward reveals wonders and possibilities you hadn't imagined before.

Are you ready to begin this extraordinary journey? Let's step through the doorway together and explore the infinite possibilities that await in the realm of parallel realities. Remember, every tremendous scientific revolution began with people being willing to question their assumptions and imagine new possibilities. You're about to join that grand tradition of explorers who dared to look beyond the obvious and discover wonders they never imagined.

What Are Parallel Realities?

I magine standing at a crossroads on a misty morning. The path splits into two directions, and you must choose one. As you stand there contemplating, a fascinating question arises: When you finally take that step in one direction, does another version of you take the other path? Does that mean you experience a completely different set of consequences and life events? This simple thought experiment brings us to the heart of one of science's most intriguing possibilities: parallel realities.

Definition and Core Concept

In its simplest form, a parallel reality is a version of our universe that exists simultaneously with our own but with some key differences. These differences might be as subtle as wearing a blue shirt instead of a red one today or as dramatic as dinosaurs never going extinct. The core idea suggests that every possible version of reality exists somewhere.

But what do we mean by "exists"? This is where the concept becomes both fascinating and complex. In science, existence isn't limited to things we can directly observe or touch. After all, we can't see radio waves. Still, we know they exist because we can detect their effects and mathematically predict their behavior. Similarly, parallel realities might exist in ways we can't directly observe but can understand through mathematics and theoretical physics.

The idea of parallel realities appears in scientific and philosophical contexts, though with essential differences. From a scientific perspective, these concepts emerge from mathematical models and quantum mechanics, suggesting that parallel universes might exist as part of reality's physical structure. Scientists focus on how these realities might arise from fundamental physical laws, seeking experimental evidence and mathematical consistency to support their theories.

Conversely, the philosophical interpretation delves into more profound questions about possibility and necessity. It explores profound questions about free will and determination, examining the relationship between consciousness and reality while contemplating the very meaning of existence itself.

Humans have been captivated by the idea of alternate realities for centuries, though in different forms. Ancient cultures often spoke of parallel worlds or realms existing alongside our own. Writers and artists have explored these concepts through science fiction and fantasy. At the same time, philosophers have long debated the nature of reality and possibility. Many religious and spiritual traditions include concepts of multiple realms or dimensions. This enduring fascination reflects something fundamental about human consciousness: our ability to imagine "what if" scenarios and contemplate alternative possibilities.

The Role of Quantum Mechanics

Quantum mechanics, developed in the early 20th century, revealed something extraordinary about the nature of reality at its most minor scales. Unlike the predictable, deterministic world we experience daily, the quantum realm operates according to principles that defy common sense. At this level, particles can exist in multiple states simultaneously, a phenomenon known as superposition. The very act of observation affects the behavior of particles, and quantum events appear to be genuinely random rather than predetermined. These discoveries suggested that reality at its most fundamental level is far stranger and more complex than we had imagined.

In 1957, physicist Hugh Everett III proposed what became known as the "many-worlds interpretation" of quantum mechanics. This theory suggests that every time a quantum event occurs with multiple possible outcomes, reality splits into different branches, each representing one possible outcome. All these branches continue to exist and evolve independently, and this process happens continuously, creating an ever-branching tree of alternate realities. This wasn't just philosophical speculation—it emerged as a serious attempt to explain the mathematical equations of quantum mechanics. While controversial, it remains one of the most discussed interpretations of quantum theory.

Two quantum phenomena particularly relevant to parallel

realities are superposition and entanglement. Superposition allows particles to exist in multiple states simultaneously, combining or "superposing" until measured. The famous "Schrödinger's cat" thought experiment illustrates this concept, suggesting that reality might exist simultaneously in multiple states. Quantum entanglement, which Einstein called "spooky action at a distance," shows how particles can become connected, sharing properties regardless of distance. This suggests a deep interconnectedness in reality and might provide clues about how parallel realities could interact.

Everyday Life and Parallel Realities

Understanding parallel realities can transform how we think about our choices and their consequences. Every decision we make might create multiple branches of reality, with each decision point representing a potential split in our life path. The fact we experience is just one of many possibilities we could be experiencing. This raises interesting questions about responsibility and meaning: if all possibilities exist, what makes our choices meaningful? The answer lies in recognizing that the reality we experience is most relevant to us, and our decisions shape which branch of reality we experience.

The concept of infinite possibilities has profound implications for how we view our lives. Mathematically, the number of possible realities might be endless, with each moment branching into countless possibilities. Every possibility, no matter how unlikely, might exist somewhere. However, it's essential to understand that not all possibilities are equally likely or accessible. Our choices still have real consequences in our experienced reality, and understanding possibilities doesn't negate responsibility. The infinity of possibilities can be both liberating and overwhelming, focusing our attention on which possibilities we want to experience and encouraging active participation in shaping our reality.

Exploring parallel realities, even as a thought exercise, can facilitate remarkable personal development. It helps us see beyond current limitations, encourages creative problem-solving, and develops flexibility in thinking. This perspective provides new ways to think about past decisions, helps process regret and "what if" scenarios, and encourages accepting life's uncertainties. Thinking about the future inspires consideration of multiple possible outcomes, encourages proactive life design, and helps identify preferred possibilities.

The concept of parallel realities challenges us to reconsider the very nature of our existence and choices. As you reflect on these ideas, consider how your life choices might differ if you consciously considered parallel possibilities. What does the possibility of parallel realities suggest about free will and determination? How might this understanding help you make better decisions? These questions invite us to delve deeper into our understanding of reality itself. The fascinating relationship between space, time, and consciousness—how these elements interweave to create the rich pattern of existence as we know it—can be understood in new ways through the lens of parallel realities.

The Multiverse: Infinite Realities

Picture yourself standing in front of a mirror in a clothing store's dressing room. The mirrors on either side create an infinite tunnel of reflections, each showing a slightly different angle of yourself stretching into a seeming infinity. This familiar experience offers us a glimpse into one of the most mind-expanding concepts in modern science: the multiverse.

What is the Multiverse?

The multiverse represents the grandest and most sweeping vision of reality ever conceived by human minds. Unlike our traditional view of a single, all-encompassing universe, the multiverse theory suggests that our universe is just one of many—perhaps infinitely many—universes existing simultaneously. Each of these universes might operate under different physical laws, contain various forms of matter and energy, or represent other possible outcomes of cosmic and mundane events.

This concept might sound like it belongs in the realm of science fiction, and indeed, popular culture has enthusiastically embraced it. Marvel's cinematic universe has brought multiverse adventures to mainstream audiences, with superheroes traversing different realities and encountering alternate versions of themselves. DC Comics has long featured a "multiverse" where different versions of their heroes exist on parallel Earths. But beneath these entertaining interpretations lies a serious scientific theory with profound implications for our understanding of reality.

The scientific foundation for the multiverse comes from several directions. Einstein's theory of relativity showed us that time and space are fundamentally connected and can bend and stretch in ways our everyday experience doesn't reveal. Quantum mechanics demonstrated that at the most minor scales, reality behaves in ways that demand the existence of multiple possibilities occurring simultaneously. String theory, while still unproven, suggests that our universe might be one of many existing on different dimensional membranes, or "branes."

Types of Multiverse Theories

As scientists have explored these possibilities, several distinct concepts of the multiverse have emerged, each with its own fascinating implications. The bubble universe theory suggests that our universe is one of many that formed during cosmic inflation—the rapid expansion that occurred just after the Big Bang. Like bubbles forming in boiling water, these universes might be continually forming, expanding, and possibly even colliding with one another in the vast expanse of space-time.

Membrane theory, arising from string theory, presents an even more exotic picture. It suggests that our entire universe might exist on a three-dimensional membrane floating in higher-dimensional space, like a sheet of paper in a three-dimensional room. Other universes might exist on their own membranes, parallel to ours but separated by dimensions we can't directly perceive. These membrane universes might occasionally interact, perhaps even collide, potentially explaining some of the mysterious phenomena we observe in our cosmos.

The quantum multiverse theory, perhaps the most mind-bending of all, emerges from quantum mechanics. It suggests that reality branches into different possibilities every time a quantum event occurs—which happens countless times every second. In one branch, a particle spins one way; in another branch, it spins the opposite way. These branches divide endlessly, creating an ever-expanding tree of alternate realities. Under this interpretation, everything that could happen happens—somewhere in the vast tapestry of the multiverse.

Why the Multiverse Matters

The implications of the multiverse theory extend far beyond academic curiosity. If true, it fundamentally changes our understanding of reality and our place within it. The multiverse suggests that the apparent fine-tuning of our universe for life—the precise values of fundamental constants that make our existence possible—might be explained by the simple fact that in an infinite number of universes, some would naturally have the right conditions for life to emerge.

This raises profound philosophical questions. If every possible version of reality exists somewhere, what does that mean for concepts like free will and moral responsibility? If somewhere there's a universe where you made different choices, does that make your choices in this universe any less meaningful? Rather than diminishing the importance of our decisions, the multiverse perspective might heighten it—after all, we can only experience and influence the reality we inhabit, even if others exist.

The multiverse concept also changes how we think about possibility and limitation. In our everyday lives, we often feel constrained by circumstances, thinking, "If only things were different." The multiverse suggests that somewhere, things are different—but it also reminds us that we must work with the reality we inhabit. This perspective can be both liberating and grounding: liberating because it helps us imagine greater possibilities and focuses on making the most of our particular slice of reality.

The ethical implications are equally fascinating. If infinite versions of ourselves exist, making every possible choice, does that mean all choices are equally valid? Most philosophers would argue no—we are responsible for our choices in our experienced reality, regardless of what other versions of ourselves might do. In fact, the multiverse perspective might increase our sense of responsibility as we realize that our choices actively select which branch of reality we experience.

Perhaps most profoundly, the multiverse theory changes our perspective on existence itself. The traditional question "Why does anything exist?" transforms into "Why do we experience this particular reality?" This shifts our focus from passive observation to active participation in reality. We're not just observers of the universe—we're participants in selecting which branch of the multiverse we experience through our choices and actions.

As we contemplate these ideas, we might feel both expanded and humbled—expanded by the infinite possibilities the multiverse suggests and humbled by our small but significant role within it. This tension between the infinite and the individual, between possibility and actuality, creates a rich framework for understanding our place in reality.

The multiverse reminds us that reality is far richer and more mysterious than we typically imagine. It suggests that the boundaries between possible and impossible might be more fluid than we thought while simultaneously grounding us in the importance of our choices in this particular reality. Whether or not we can ever definitively prove the existence of the multiverse, contemplating it enriches our understanding of existence and opens our minds to greater possibilities.

These abstract concepts of multiple realities might manifest in our everyday experience, and understanding them can enhance our approach to life's challenges and opportunities. The multiverse isn't just a distant scientific concept but a powerful lens through which to view our own existence and potential.

Time Travel and Its Connection to Alternate Realities

The steady ticking of a clock marks our journey through time, always moving forward, never backward. Or so we think. Yet, where parallel realities and quantum physics intersect, time is far more complex and mysterious than our everyday experience suggests. The possibility of time travel, long a staple of science fiction, has emerged as a serious topic of scientific inquiry, deeply intertwined with our understanding of alternate realities.

What Time Travel Could Mean for Reality

When we speak of time travel, we must first distinguish between the familiar forward march of time we all experience and the more exotic possibilities that physics suggests exist. Every day, we travel through time at one second per second. Still, Einstein's theory of relativity reveals that this rate is more flexible than we might think. Time can stretch and compress depending on how fast we're moving or how close we are to massive objects—a phenomenon already confirmed by precise atomic clocks in satellites orbiting Earth.

But what about more dramatic forms of time travel? The same equations that Einstein used to describe space-time suggest the theoretical possibility of loops in time—pathways that could return to the past or future at rates far beyond our everyday experience. These aren't just mathematical curiosities; they point to profound questions about the nature of reality itself.

The grandfather paradox, the most famous thought experiment in time travel, illustrates the deep connection between time travel and parallel realities. Imagine traveling back in time and preventing your own grandparents from meeting. This would prevent your own birth, making it impossible for you to have traveled back in time in the first place. The paradox seems unsolvable—unless we consider the possibility of branching timelines and alternate realities.

Another mind-bending concept is the bootstrap paradox, where information or objects seem to have no original source. Imagine finding plans for a time machine, using them to build one, and then traveling back in time to give yourself those same plans. Where did the information initially come from? These paradoxes suggest that if time travel is possible, reality must be structured in ways far more complex than our linear perception suggests.

The Role of Time in Alternate Realities

In parallel realities, time might flow differently than in our own universe. Some theories suggest that time might be relative to motion and gravity but to the particular reality you're experiencing. This leads to the fascinating possibility that different versions of you might be experiencing different moments simultaneously—or that the concept of "simultaneous" might need to be reconsidered when dealing with multiple realities.

The block universe theory takes this even further, suggesting that the past, present, and future exist simultaneously, like different locations in space. Just as New York and Tokyo exist simultaneously, even though we can only be in one place at once, this theory suggests that yesterday and tomorrow exist "now," even though we can only experience one moment at a time. This view of time as a dimension rather than a flow has profound implications for understanding time travel and parallel realities.

The butterfly effect—the idea that tiny changes can have enormous consequences—becomes particularly relevant when considering time travel and alternate realities. Each small decision or change might spawn new timelines in a multiverse where all possibilities exist, creating an ever-branching tree of alternate realities. A single choice, like stopping to tie your shoelace, might make one timeline where you miss your bus and another where you catch it, each leading to dramatically different futures.

Time Travel in Science and Culture

Our fascination with time travel has spawned countless stories, from H.G. Wells's "The Time Machine" to modern films like "Interstellar" and "Back to the Future." These narratives often grapple with the personal and ethical implications of changing the past. What would you do if you could go back and change a crucial moment in your life? Would it be correct to prevent historical tragedies? These stories resonate because they touch on fundamental human desires—to correct mistakes, understand consequences, and see what might have been.

While we haven't built any time machines that can take us to the past, scientists have created remarkable experiments that reveal time's malleability. Particle accelerators regularly create particles that experience time differently than we do, living far longer than they would if stationary. GPS satellites must constantly adjust their clocks to account for time dilation effects predicted by Einstein's theories. These real-world experiments show that time isn't as rigid as our everyday experience suggests.

The philosophical implications of time travel extend far beyond simple cause and effect. If the past can be changed, what does that mean for free will and moral responsibility? If multiple timelines exist, which one is "real"? Perhaps all of them are, suggesting that reality is far richer and more complex than we typically imagine.

Time travel also raises questions about the nature of memory and identity. If you travel to the past and change events, would your memories instantly update to reflect the new timeline, or would you remember the "original" sequence? Some theories suggest that consciousness might play a crucial role in how we experience the flow of time and navigate between possible realities.

The intersection of time travel and parallel realities suggests that our universe might be even stranger than we imagine. We think of the fixed past as just one branch of an infinite tree of possibilities. Every moment of decision creates new branches, timelines, and realities. This perspective challenges our traditional understanding of cause and effect, suggesting that reality might be more like a vast network of interconnected possibilities than a simple linear progression.

Understanding these concepts can change how we view our lives and choices. While we might not be able to physically travel through time, recognizing the complex relationship between time, choice, and reality can help us appreciate the significance of each moment. Every decision we make might create new branches of reality, making us not just participants in reality but active creators of new possibilities.

As we explore parallel realities, we'll see how these ideas about time and possibility relate to quantum mechanics and the structure of the universe itself. The nature of time—whether linear or branching, fixed or flexible—remains one of science's greatest mysteries, intimately connected to our understanding of parallel realities and the fundamental nature of existence.

The quantum mechanical principles that underlie these fascinating possibilities reveal how the bizarre behavior of subatomic particles might hold the key to understanding both time travel and parallel realities. The minor scales of reality help explain its most significant mysteries and quantum mechanics suggests that reality is stranger than time travel stories imagine.

Alternate Dimensions: Beyond Our 3D World

Imagine an ant walking on a piece of paper. From its perspective, the world consists only of forward, backward, left, and proper—a perfect two-dimensional existence. Now picture lifting that ant off the paper with your finger. This movement would seem like magic to the ant, a mysterious force pulling it in an impossible direction. This simple thought experiment opens the door to understanding reality's most fascinating aspects: the possibility of dimensions beyond the ones we can see.

Beyond the Third Dimension

We live in a world that seems perfectly content with three spatial dimensions: height, width, and depth. Every object we encounter, from the smallest grain of sand to the most prominent mountain, can be described using these three coordinates. We move through these dimensions effortlessly, barely considering how fundamental they are to our reality experience. Yet mathematics and physics suggest that three dimensions might be just the beginning.

A dimension, at its most basic, is simply a direction of movement or measurement. When you walk down a street, you're moving through one dimension. When you climb stairs while walking forward, you move through two dimensions simultaneously. Adding a sideways component makes it three. But why stop there? Mathematics is fine with describing spaces with four, five, or even infinite dimensions. Each additional dimension represents a new direction of movement, a new axis along which reality might extend.

Time itself can be considered a fourth dimension, though it behaves differently from spatial dimensions. Einstein's theory of relativity reveals that time and space are intimately connected, forming what physicists call spacetime. Just as we can move through space in three directions, we move through time in one direction—at least in our everyday experience. This perspective transforms time from a mere background against which events occur into an actual dimension of reality, as natural and fundamental as height, width, and depth.

String Theory and the Hidden Dimensions

String theory, one of physics' most ambitious attempts to explain the fundamental nature of reality, suggests something remarkable: our universe might contain many more dimensions than we can see. According to this theory, the basic building blocks of reality aren't point-like particles but tiny vibrating strings. These strings, however, require additional dimensions to vibrate in ways that could produce all the particles and forces we observe.

How many extra dimensions? Different versions of string theory suggest anywhere from six to twenty-two additional spatial dimensions. These dimensions wouldn't be spread out like our familiar three but would be "compactified"—curled up so tightly that they're invisible to our direct observation. Imagine an ant walking along a tightrope. From far away, the rope appears to be one-dimensional, just a line. But up close, the ant can walk around the rope's circumference, experiencing a second dimension that wasn't visible from a distance.

Similarly, these extra dimensions might be curled up at every point in space, which is too small to see but still influences how reality works. They could explain mysterious aspects of our universe, from the strength of gravity to the existence of dark matter. These hidden dimensions might even contain entire universes, separated from our own by the tiniest of distances but in directions we can't perceive.

Visualizing the Impossible

How can we possibly understand dimensions we can't see or directly experience? This is where analogies become potent tools. The classic book "Flatland" by Edwin Abbott, Abbott describes a two-dimensional world inhabited by geometric shapes. These flat beings can't conceive of "up" or "down" until they're visited by a three-dimensional sphere, which appears to them as a circle that can magically change size as it passes through their plane of existence.

We can extend this analogy to understand higher dimensions.

Just as a three-dimensional object passing through a two-dimensional plane would change shape mysteriously, a four-dimensional object passing through our three-dimensional space would alter its three-dimensional shape in seemingly impossible ways. A four-dimensional sphere called a hyper-sphere, would appear to us first as a point, then as a growing sphere, then as a shrinking sphere, before disappearing—just as a regular sphere passing through a plane appears to a two-dimensional being as a growing and shrinking circle.

The implications of higher dimensions extend far beyond mathematical curiosities. They might explain fundamental aspects of our universe that currently puzzle scientists. Gravity, for instance, seems unusually weak compared to other fundamental forces. String theory suggests this might be because gravity spreads through all dimensions while other troops are confined to our familiar three dimensions, making gravity appear weaker than it really is.

Higher dimensions also provide new ways to think about parallel realities. Suppose our universe is like a three-dimensional sheet floating in a higher-dimensional space. In that case, other universe-sheets might exist parallel to ours, separated by distances in dimensions we can't perceive. This "braneworld" scenario suggests that what we think of as parallel universes might be more like parallel planes of existence, literally stacked in higher dimensions.

The technological implications of understanding higher dimensions are equally fascinating. Some theoretical physicists suggest that certain quantum computing operations might effectively use higher dimensions to perform calculations impossible in ordinary three-dimensional space. Artificial intelligence and machine learning engineers already use high-dimensional mathematics to model complex systems and patterns.

Perhaps most intriguingly, higher dimensions might help explain consciousness itself. Some theories propose that conscious experience involves processing information across multiple dimensions beyond the three we can see. The fact that we can imagine and mathematically describe higher dimensions, even though we can't directly perceive them, is evidence of our minds' connection to higher-dimensional reality.

We might feel overwhelmed and exhilarated as we grapple with these concepts. The possibility of dimensions beyond our direct experience reminds us that reality is more decadent and mysterious than our everyday perception suggests. Yet our ability to understand and work with these concepts, even if only mathematically, shows the remarkable capacity of the human mind to reach beyond its apparent limitations.

These ideas of higher dimensions connect to quantum mechanics and the nature of consciousness, suggesting that the structure of reality might be even more intricate than dimensions alone can describe. The mathematics of higher dimensions might help explain the physical universe and our experience of it as conscious beings.

Scientific Experiments Supporting Parallel Realities

In a dimly lit laboratory, a scientist sends individual particles of light through two narrow slits. The resulting pattern on the detector screen tells a story so strange, so contrary to our everyday experience, that it forces us to question the very nature of reality. This famous double-slit experiment is perhaps the most dramatic demonstration that our universe behaves in ways that demand the existence of parallel realities.

Quantum Experiments and Parallel Worlds

The double-slit experiment reveals something remarkable: a single particle, fired alone, somehow seems to travel through both slits simultaneously, interfering with itself like a wave. When scientists try to observe which slit the particle actually travels through, the interference pattern disappears, and the particle behaves like an ordinary particle again. It's as if observation forces reality to choose one path from many possible paths—or perhaps forces us to experience just one of many parallel realities.

This phenomenon lies at the heart of quantum mechanics' most profound mystery: superposition. A quantum system can exist in multiple states simultaneously until it's observed or measured. Erwin Schrödinger famously illustrated this concept with his thought experiment about a cat in a box, neither alive nor dead, but existing in a superposition of both states until someone opens the box to check. While this was meant to highlight the absurdity of applying quantum principles to everyday objects, it raises a fascinating possibility: perhaps the cat is alive and dead, with each possibility existing in a different parallel reality.

Even more mysterious is quantum entanglement, what Einstein called "spooky action at a distance." When particles become entangled, measuring one instantly affects the other, no matter how far apart. Recent experiments have demonstrated this effect over distances of hundreds of kilometers. Some theorists suggest that this instantaneous connection might be explained by parallel realities—the particles aren't communicating faster than light; instead, they're connected through higher dimensions or parallel universes.

Modern quantum experiments have become increasingly sophisticated. The "quantum eraser" experiment shows that a measurement made in the present can affect what happened in the past. Quantum tunneling experiments demonstrate particles appearing to traverse barriers instantly as if taking shortcuts through other dimensions. These experiments add another piece to the puzzle, suggesting that reality is far stranger than our classical intuition leads us to believe.

Cosmic Evidence for the Multiverse

The quest for evidence of parallel realities extends far beyond quantum laboratories to the vast scales of the cosmos. The cosmic microwave background radiation—the afterglow of the Big Bang—contains patterns that some scientists interpret as possible evidence of collisions with other universes. These "bruises" in the early universe might indicate that our cosmos is one bubble in a vast foam of universes, each with its own physical laws and constants.

Gravitational waves, first directly detected in 2015, have opened a new window into the universe. These ripples in spacetime might help us detect evidence of other universes. Some theories suggest that if parallel universes exist as "branes" floating in higher dimensions, their collisions might generate distinctive gravitational wave patterns we could detect.

The mysterious dark matter and energy that comprise most of our universe's content might also connect to parallel realities. Some theorists propose that dark matter's gravitational effects could be explained by matter existing in parallel dimensions, affecting our universe only through gravity. Dark energy, driving the universe's accelerating expansion, might represent forces leaking in from other dimensions or universes.

The distribution of galaxies in the universe, the apparent fine-tuning of physical constants, and even certain anomalies in cosmic ray observations have all been cited as possible evidence for a multiverse. While none of these observations provides definitive proof, they create a compelling pattern that many scientists find difficult to explain without invoking parallel realities.

Philosophical Implications of the Evidence

The experimental evidence for parallel realities forces us to confront deep philosophical questions about the nature of existence and observation. The Copenhagen interpretation of quantum mechanics—long the standard view—suggests that reality remains indefinite until observed. But this raises profound questions: What constitutes an observation? Does consciousness play a unique role in determining reality?

The many-worlds interpretation offers a different perspective, suggesting that all possible outcomes occur in various branches of reality. This view eliminates the unique role of observation but requires us to accept an ever-branching universe of infinite possibilities. These aren't just abstract considerations—they affect how we interpret every quantum experiment and cosmic observation.

Science and philosophy work together here in ways rarely seen in other fields. The philosophical implications of quantum experiments help guide which new experiments might be most revealing. At the same time, experimental results force philosophers to refine and sometimes radically revise their concepts of reality. This interplay has led to new interpretations that attempt to bridge the gap between quantum and classical reality, such as quantum decoherence theory.

Why should the average person care about these esoteric experiments and their interpretations? Because they challenge our most basic assumptions about reality, causality, and free will. If parallel realities exist, every choice we make might spawn new universes. If consciousness plays a role in determining reality, our observations might be more fundamental to the universe than we imagined. These ideas have practical implications for our view of decision-making, responsibility, and the nature of possibility.

The experiments also highlight the limitations and strengths of human perception and understanding. We've developed tools and mathematics to probe reality far beyond our direct sensory experience, revealing a universe—or multiverse—far stranger and more wonderful than we could have imagined. This should inspire humility about our current understanding and excitement about what we might discover.

Modern experiments continue to push the boundaries of our understanding. Quantum computers, now becoming a reality, manipulate superpositions of quantum states in ways that might be explained by parallel realities processing information simultaneously. Experiments with time and causality at the quantum level challenge our basic notions of cause and effect. Each new discovery suggests that reality is stranger than we suppose but stranger than we can think.

As we move forward, new experiments will undoubtedly reveal more surprises. Plans are underway for larger quantum computers, more sensitive gravitational wave detectors, and more precise cosmic microwave background measurements.

Each of these might provide new evidence for or against parallel realities. What's certain is that we're living in an extraordinary time when questions once confined to philosophy are becoming accessible to experimental science.

These scientific discoveries may connect deeply to human consciousness and experience, raising questions about whether our minds might be uniquely equipped to understand—or even navigate—a reality that includes parallel worlds. The evidence suggests we're part of something far more extensive and mysterious than we once imagined, and we're only beginning to understand its true nature.

Philosophical Implications of Parallel Realities

W hat does it mean to choose in a universe where all possibilities occur? If somewhere, in some parallel reality, you're already living every possible version of your life, does anything you do matter? These questions might seem purely academic, but they strike at the heart of how we understand ourselves and our place in existence. The concept of parallel realities doesn't just challenge our scientific understanding—it forces us to reconsider fundamental questions about free will, identity, and the very nature of being.

Free Will in an Infinite Multiverse

Imagine standing at a crossroads, deliberating between two paths. Traditional philosophy already needs to work on whether this choice is truly free or predetermined by the chain of cause and effect. However, parallel reality theory adds another layer of complexity: perhaps both choices occur, each in its own branch of reality. Does this mean your deliberation is meaningless since both options will be taken? Or does it make your choice even more significant, as you're actively selecting which branch of reality you'll experience?

The determinism versus free will debate takes on new dimensions in a multiverse context. Classical determinism suggests that all events inevitably follow from prior causes, leaving no room for actual choice. However, determinism might apply differently in a multiverse where all possibilities occur. What's determined is not which choice you'll make but rather the full spectrum of options that will branch into different realities. Your free will might lie not in changing what's possible—since everything possible happens somewhere—but in navigating which possibility you experience.

This perspective has fascinated philosophers for centuries, even before modern physics suggested its literal truth. Leibniz spoke of "possible worlds" where different versions of reality played out. Now, we must grapple with the possibility that these aren't just philosophical thought experiments but actual realities. The ancient question "Could I have done otherwise?" transforms into "Am I doing otherwise, in another reality?"

Existence Across Multiple Timelines

Existentialism traditionally focuses on the individual's struggle to find meaning in a seemingly meaningless universe. But how does this change when we consider existence across multiple realities? The existentialist emphasis on personal choice and authenticity takes on new significance when every choice creates new branches of reality. Perhaps authenticity lies not in making the "right" choice but in consciously choosing which reality we want to experience.

The question of personal identity becomes particularly challenging in this context. Are you the same person across all these different realities? If consciousness somehow spans multiple timelines, what does that mean for our sense of self? Some philosophers suggest that our identity might be more like a vast tree of possibilities than a single linear narrative. Each choice we make might be less about determining our future and more about selecting which branch of our already-existing multiple selves we align with.

This perspective challenges traditional existentialist ideas about anxiety and responsibility. The existential dread of facing unlimited possibilities might be even more acute if we consider that all possibilities actually occur. Yet this same infinity of options might also be liberating. Somewhere, every version of success is actually happening. The challenge becomes not what to make of ourselves in a single reality but how to navigate our journey through an infinity of possible selves.

The Ethics of Multiple Realities

The ethical implications of parallel realities are the most challenging to grapple with. If every possible choice occurs somewhere, does that mean all options are equally valid? Most philosophers would argue no—the fact that something happens somewhere doesn't make it right. Our ethical responsibility might lie in consciously choosing which realities we want to bring into our experienced existence.

Consider the question of moral responsibility across realities.

If a version of you makes every possible choice somewhere, are you responsible for the actions of these alternate selves? This raises profound questions about the nature of moral agency and responsibility. Perhaps our ethical duty is not to prevent bad choices from occurring—since they appear somewhere in the multiverse anyway—but to actively choose to experience and create realities aligned with our highest values.

The existence of parallel realities suggests the existence of parallel moral frameworks. In some realities, entirely different ethical systems might have evolved. This doesn't necessarily lead to moral relativism—the fact that other moral systems exist wouldn't make them all equally valid. Instead, it might point to a more profound universal ethics that transcends individual realities, governing how we should navigate between possible worlds.

In this view, our place in the universe becomes both more significant and more humble. We're more humble because we're just one version of ourselves among infinite possibilities, and our choices actively select and create the reality we experience. We might be less the authors of reality than its editors, choosing which version of the infinite story we want to experience and help create.

These philosophical implications extend into practical questions about how to live. If parallel realities exist, should we spend more time considering alternate possibilities? Should we be more or less careful about our choices, knowing that all possibilities play out somewhere? Perhaps the answer lies in developing a new kind of wisdom—not just about what to choose, but how to determine which reality to experience.

The concept of parallel realities also raises profound questions about the nature of regret and possibility. If every choice we didn't make is taken by another version of ourselves, perhaps regret becomes less about what we didn't do and more about understanding why we chose the reality we did. This might lead to a more conscious and intentional way of making choices, aware that we're not just deciding what to do but actively selecting which reality to align ourselves with.

The relationship between consciousness and reality takes on new significance in this context. Suppose consciousness plays a role in collapsing quantum possibilities into experienced reality, as some interpretations of quantum mechanics suggest. In that case, our awareness might be more fundamental to the structure of reality than we imagined. We might be not just observers but active participants in selecting which reality becomes "real" for us.

As we look forward, these philosophical implications suggest new ways of thinking about human potential and purpose. Our role in the universe is not just to exist but to consciously navigate the infinite possibilities. This might require developing new philosophical frameworks that can handle the complexity of multiple realities while still providing practical guidance for living.

These philosophical implications invite us to reconsider abstract concepts and how we approach our daily lives and choices. Understanding parallel realities might offer more than intellectual satisfaction—it could provide a fundamental shift in how we view our role in the cosmos. As we continue to explore and understand these concepts, we might find that parallel realities are not merely a scientific curiosity but rather a key to understanding our unique place in the boundless pattern of reality, offering new perspectives on how we make decisions, form relationships, and create meaning in our lives.

Pop Culture and Parallel Realities

E ntertainment media has played a pivotal role in shaping our understanding of parallel realities. Through groundbreaking films like "The Matrix," audiences have struggled with profound questions about the nature of reality and consciousness. The film's exploration of simulation theory has sparked countless discussions about whether our reality might be a constructed experience. The iconic "red pill or blue pill" scene has become a cultural touchstone, representing the choice between comfortable illusion and complicated truth. This metaphor has transcended the film, entering everyday discourse about truth, reality, and perception.

Similarly, Marvel's expansive multiverse has brought complex scientific concepts into mainstream conversation, using alternate timelines and parallel worlds to weave intricate narratives that capture imaginations worldwide. Films like "Doctor Strange" and "Spider-Man: Into the Spider-Verse" have visualized these concepts in stunning ways, making abstract theories accessible through compelling storytelling and groundbreaking animation. Christopher Nolan's "Interstellar" took a more grounded approach, using established physics to explore how time might bend and twist in ways we're only beginning to understand. The film's emotional core—a father's relationship with his daughter across time and space—demonstrates how parallel reality narratives can illuminate deeply human stories.

Literature has long been at the forefront of exploring alternate realities. Philip K. Dick's works stand as masterpieces of the genre, consistently challenging readers to question the nature of reality and consciousness. His novel "The Man in the High Castle" presents an alternate history where the Axis powers won World War II, demonstrating how parallel world narratives can illuminate our reality through contrast. The novel's structure plays with reality, featuring a book-within-a-book that depicts an alternate timeline closer to our history. Dick's other works, like "Ubik" and "Flow My Tears, the Policeman Said," continue to influence modern storytelling with their reality-bending narratives. Science fiction literature bridges complex scientific theories and public understanding, making abstract concepts accessible through compelling storytelling. Authors like Jorge Luis Borges, with "The Garden of Forking Paths," and Michael Moorcock, with his Eternal Champion series, have expanded the literary possibilities of parallel reality narratives.

The influence of these pop culture explorations extends far beyond entertainment. When audiences engage with parallel reality narratives, they often process deep philosophical questions about free will, destiny, and the nature of existence itself. Television series like "Fringe" have explored these concepts over multiple seasons, developing complex mythologies around parallel worlds and examining how small changes in history might create drastically different outcomes. The series "Dark" on Netflix took this concept further, weaving an intricate tale of time travel and parallel worlds that challenge viewers to consider the interconnectedness of all things and the profound implications of causality.

These stories resonate because they tap into fundamental human curiosities: What if I had made different choices? What other versions of reality might exist? What defines who we indeed are? Through creative storytelling, these abstract concepts become tangible and relatable, allowing us to explore complex ideas through the lens of human experience. Shows like "Black Mirror" have demonstrated how technological advancement might create new forms of parallel existence. At the same time, "Rick and Morty" uses humor and absurdity to explore the philosophical implications of infinite realities.

Media portrayals of parallel realities reflect and shape our collective imagination, influencing how we conceptualize the possibilities of existence. Video games have added an interactive dimension to these explorations, with titles like "BioShock Infinite" and "Quantum Break" allowing players to actively engage with parallel reality narratives. These interactive experiences create a deeper understanding of cause and effect across multiple timelines while raising questions about choice, consequence, and the nature of reality itself.

The artistic expression of parallel realities has evolved alongside our scientific understanding. As quantum physics reveals more about the potentially probabilistic nature of reality, creators have found new ways to express these concepts through various media. Music videos, virtual reality experiences, and interactive art installations contributed to our cultural understanding of parallel realities. Artists like David Lynch have pushed the boundaries of conventional narrative, creating works that seem to exist in multiple realities simultaneously.

The enduring popularity of parallel reality narratives speaks to something fundamental in human nature—our desire to explore the unknown and understand our place in an increasingly complex universe. These stories provide a framework for contemplating possibilities beyond our immediate experience, encouraging us to think more deeply about the nature of reality and consciousness. They help us process complex emotions about choice and regret, offering catharsis by exploring "what-if" scenarios.

The impact of these narratives extends into everyday life, influencing how people think about decision-making and personal identity. The concept of parallel realities has become a valuable tool for mental health professionals, helping clients process regret and anxiety about life choices. It has also influenced fields as diverse as business strategy, where scenario planning often involves considering multiple possible futures, and education, where understanding various perspectives and possibilities is increasingly valued.

As science continues to reveal the strange nature of our universe, pop culture serves as a vital tool for processing and understanding these revelations, helping us navigate the fascinating intersection of imagination and scientific discovery. The stories we tell about parallel realities reflect our growing awareness of the universe's complexity and our place within it. Through these narratives, we find ways to explore the most profound questions about existence, consciousness, and the nature of reality itself.

The Future of Parallel Realities Research

The frontier of parallel realities research stands at a fascinating intersection of quantum mechanics, cosmology, and cutting-edge technology. At CERN's Large Hadron Collider, scientists are probing the fundamental nature of reality through increasingly sophisticated particle collisions. These experiments generate conditions similar to those present in the earliest moments of our universe, potentially revealing evidence of extra dimensions or parallel worlds. Recent upgrades to the collider have enabled the detection of more subtle quantum phenomena, pushing our understanding of particle physics to new limits. Each collision event generates massive amounts of data that, when analyzed, might hold clues to the existence of parallel realities or hidden dimensions.

Quantum mechanics continues to yield surprising discoveries that challenge our conventional understanding of reality. Experiments in quantum entanglement have demonstrated what Einstein famously called "spooky action at a distance," suggesting connections between particles that transcend our traditional concepts of space and time. Scientists have successfully maintained quantum coherence at increasingly large scales, bringing quantum effects into the macroscopic world. These achievements raise intriguing possibilities about the nature of reality and consciousness, suggesting that quantum phenomena play a role in biological systems and even human cognition.

Modern telescopes and observational technologies are revolutionizing our view of the cosmos. The James Webb Space Telescope and other advanced instruments provide unprecedented glimpses into the early universe, potentially revealing signatures of other universes or dimensions that might have influenced our cosmic evolution. Gravitational wave detectors have opened a new window into the universe, allowing scientists to detect ripples in spacetime. These observations might eventually reveal evidence of collisions between our universe and others or provide clues about the existence of parallel realities.

Quantum computers could simulate aspects of parallel realities that are impossible to study through conventional means, thanks to their ability to manipulate quantum states directly. As these technologies advance, they may enable new approaches to understanding the multiverse.

Integrating virtual and augmented reality technologies with scientific research creates new ways to visualize and interact with complex theoretical models. Scientists can now "walk through" simulated multiple dimensions or observe quantum phenomena at human scales. These visualization tools are not just educational aids but genuine research instruments, helping theorists develop new insights into the nature of reality.

As we look ahead to the next fifty years, parallel realities research promises revolutionary discoveries. Advanced quantum sensors detect interactions between parallel universes. At the same time, new mathematical frameworks could provide testable predictions about the structure of the multiverse. The development of more sophisticated quantum computers could allow us to simulate and study parallel realities in unprecedented detail, potentially bridging the gap between theoretical predictions and experimental evidence.

The next generation of scientists will likely approach these questions with tools and perspectives we can barely imagine today. The convergence of quantum computing, artificial intelligence, and advanced sensing technologies might reveal aspects of reality that currently seem beyond our reach. Young researchers are already developing novel experimental approaches that challenge traditional boundaries between different scientific disciplines, creating new frameworks for understanding parallel realities.

As our understanding deepens, the implications extend far beyond pure science. Studying parallel realities could revolutionize our approach to technology, leading to new forms of quantum communication or computing that harness the properties of multiple dimensions. Medical research might benefit from insights into quantum biology, potentially revealing new therapeutic approaches based on our understanding of parallel possibilities at the quantum level.

The ethical and philosophical implications of these advances are equally profound. As we develop better tools for understanding parallel realities, we must contend with questions about the nature of consciousness, free will, and human identity. The possibility of accessing or influencing parallel realities raises complex ethical considerations that future generations must address.

The significance of this research cannot be overstated. Understanding parallel realities isn't just about satisfying scientific curiosity—it's about comprehending the fundamental nature of existence itself. Each breakthrough in this field brings us closer to answering humanity's most profound questions about our place in the cosmos and the true nature of reality.

The current pace of scientific advancement suggests we are on the cusp of transformative discoveries. New experimental techniques are constantly being developed while theoretical frameworks evolve and mature. Combining improved observational tools, advanced computational capabilities, and refined theoretical models positions us to make unprecedented progress in understanding the nature of parallel realities.

As we explore these frontiers, the boundaries between what seems possible and impossible continue to shift. The study of parallel realities remains one of humanity's most ambitious intellectual endeavors, promising to reshape our understanding of existence and place within an increasingly complex and fascinating universe.

Personal Growth Through Exploring Parallel Realities

U nderstanding parallel realities offers more than scientific or philosophical insights—it provides a robust framework for personal transformation and growth. Our perspective on our lives fundamentally shifts when we begin to grasp the vastness of possibility inherent in the concept of parallel realities. The limitations we perceive in our daily existence start to appear less absolute, and the boundaries of what we believe possible expand.

Consider how understanding parallel realities challenges our basic assumptions about life. Every decision you've made, every path you've chosen, represents just one of countless possibilities. Rather than breeding regret, this realization can foster a profound sense of empowerment. Your current reality isn't a fixed, unchangeable state but one manifestation of infinite possibilities. This perspective can revolutionize how you approach challenges, decisions, and personal growth.

The concept of parallel realities offers a unique tool for breaking through mental and emotional blocks. When you feel stuck in a particular pattern or mindset, imagining alternate versions of yourself who made different choices can illuminate new possibilities. This isn't mere daydreaming—it's a practical exercise in expanding your perception of what's possible. By considering how alternate versions of yourself might have handled situations differently, you gain insights into your potential for change and growth.

This mindset naturally fosters creativity and innovation. When you understand that reality might be more fluid and multifaceted than we typically assume, creative barriers dissolve. Artists, writers, and innovators who embrace multiverse thinking often find themselves capable of imagining and creating things they never thought possible. The concept serves as a reminder that the boundaries of what's possible are frequently self-imposed.

Exploring parallel realities can also help heal emotional wounds and process past decisions. Instead of being haunted by "what if" scenarios, you can view them as valid alternate realities alongside your current experience. This perspective can help release attachment to past choices while honoring the validity of your journey. It transforms regret into acceptance and curiosity about the path you're currently traveling.

The practice of contemplating parallel realities naturally cultivates mindfulness and presence. When you truly understand that your current reality has infinite possibilities, each moment becomes more precious and meaningful. This awareness can deepen your appreciation for the present while maintaining an open mind about future opportunities.

This exploration can also enhance your relationships and empathy. Understanding that others are experiencing their unique reality helps develop compassion and understanding of different perspectives. It becomes easier to see how someone else's choices and beliefs make sense within their context, even if they differ from yours.

For personal development, the concept of parallel realities offers practical tools for goal-setting and achievement. Instead of viewing your desired future as a distant possibility, you can imagine it as an existing reality you're moving toward. This shift in perspective can make your goals more achievable and motivate you to take concrete steps toward them.

Maintaining an open mind cannot be overstated in this context. An open mind becomes not just a philosophical stance but a practical tool for growth. When you truly embrace the possibility of parallel realities, you naturally become more receptive to new ideas, experiences, and ways of being. This openness creates a feedback loop of growth and discovery, where each new understanding further expands consciousness.

Curiosity becomes a natural companion on this journey. As you explore these concepts, you may ask more profound questions about the nature of reality, consciousness, and your existence. These questions, rather than being merely academic, become gateways to profound personal insights and transformation.

Exploring parallel realities can also lead to a deeper understanding of one's own identity. When you consider that multiple versions of yourself might exist, it raises fascinating questions about what makes you uniquely you. This contemplation can strengthen one's sense of self while paradoxically making one more open to change and growth.

Applying these concepts to daily life involves practical exercises and mindset shifts. Consider regularly considering alternate perspectives on situations, deliberately challenging your assumptions about what's possible, or using visualization techniques incorporating parallel possibilities. These practices help integrate this expansive understanding into your daily experience.

For spiritual growth, parallel realities align with many ancient wisdom traditions that speak of multiple dimensions or planes of existence. This modern scientific framework can provide a new context for understanding traditional spiritual concepts, potentially deepening your scientific and spiritual understanding.

Exploring parallel realities ultimately leads to a more expansive and empowering view of life. It teaches us that reality is more magnificent and full of possibility than we typically imagine. This understanding can transform how we approach challenges, relationships, and personal growth, leading to a richer and more meaningful life experience.

As you continue to explore these concepts, remember that the goal isn't to escape your current reality but to enhance your experience. Understanding parallel realities should deepen your appreciation for the unique path you're on while maintaining awareness of the vast possibilities surrounding you. This balance between acceptance and possibility creates the perfect conditions for profound personal transformation and growth.

Living in a World of Infinite Possibilities

The concept of parallel realities fundamentally transforms how we view ourselves and our place in the universe. When we truly internalize the idea that infinite possibilities exist, our sense of self expands beyond the singular narrative we typically maintain. We begin to see ourselves not as fixed entities bound by past decisions but as dynamic beings existing within a vast web of potential. This shift in perspective can profoundly impact how we approach life's challenges and opportunities.

Understanding parallel realities challenges our traditional notions of identity. Rather than viewing ourselves as the product of a single timeline of choices and experiences, we can recognize ourselves as multifaceted beings capable of infinite variation. This understanding is consistent with our current knowledge. Instead, it enriches it by acknowledging the complexity and potential inherent in every moment. Our decisions become less about right or wrong and more about exploring different aspects of our potential.

Through this lens, the role of uncertainty in our lives takes on new meaning. What once might have felt threatening or destabilizing becomes an invitation to explore and grow. Uncertainty transforms from an obstacle into a gateway to possibility. This shift doesn't just change how we think about uncertainty—it changes how we live with it. We welcome the unknown as a natural part of existence, understanding that it holds the seeds of infinite potential.

Curiosity emerges as a crucial tool for navigating this expanded view of reality. Maintaining an actively curious mindset opens us to new possibilities and perspectives that might otherwise remain hidden. This isn't just about asking questions—it's about maintaining a state of wonder about existence itself. Each moment becomes an opportunity for discovery, whether exploring new ideas, meeting new people, or examining our thoughts and beliefs.

The practical application of curiosity in our daily lives can take many forms. It might mean asking more profound questions about our habitual responses to situations, exploring new approaches to familiar challenges, or seeking perspectives that differ from ours. This curious approach to life naturally leads to growth and transformation as we constantly expand our understanding of what's possible.

Living aware of infinite possibilities doesn't mean we become paralyzed by choices or lost in daydreams about alternate realities. Instead, it empowers us to move through life with greater intention and awareness. We understand that each option is meaningful, not because it's the only possible choice, but because it's the one we're exploring now.

This understanding can revolutionize how we approach personal growth and development. Rather than seeing ourselves as works in progress moving toward a fixed goal, we can embrace the idea that we're constantly exploring different facets of our potential. This perspective removes the pressure of perfection and replaces it with an excitement about discovery and experimentation.

The concept of infinite possibilities also transforms how we view our relationships and connections with others. It becomes easier to practice empathy and understanding when we understand that each person exists within their own web of possibilities. We recognize that others' choices and perspectives are valid expressions of their journey through the infinite landscape of potential.

In practical terms, living with an awareness of infinite possibilities means developing new habits of thought and action. It involves regularly questioning our assumptions, remaining open to unexpected opportunities, and maintaining flexibility in our approaches to life. This mindset allows us to navigate challenges with greater resilience and creativity, knowing there are always multiple paths forward.

The power of this perspective lies in its ability to liberate us from self-imposed limitations. When we genuinely understand that reality is more expansive than we typically imagine, we naturally question other boundaries we've accepted without examination. This questioning can lead to breakthrough insights and transformative changes in our lives and work.

Embracing the unknown becomes a central practice in this approach to life. Rather than seeking to control or predict everything, we learn to confidently move forward while maintaining openness to possibility. This balance between intention and receptivity creates optimal conditions for growth and discovery.

The impact of living with an awareness of infinite possibilities extends beyond individual growth to influence how we participate in the larger world. We become more willing to engage with complex challenges, more capable of imagining innovative solutions, and more committed to contributing to positive change. This expanded perspective naturally leads to more conscious and constructive engagement with life.

Living in a world of infinite possibilities requires ongoing commitment and attention. It's not enough to intellectually understand these concepts—we must actively integrate them into our daily experience. This means regularly examining our assumptions, maintaining an attitude of openness and curiosity, and consciously choosing how we engage with the possibilities before us.

As we cultivate this way of being, we discover that the universe is more magnificent and full of potential than we ever imagined. Our role within it becomes both more humble and more significant—humble because we recognize the vastness of possibility and essential because we understand our power to consciously participate in the unfolding of reality.

This understanding leads us to approach life with incredible wonder, engagement, and purpose. We recognize that each moment holds infinite potential, and our choices help shape the possibilities that manifest in our experiences. This awareness infuses everyday life with more profound meaning and excitement as we consciously participate in the ongoing creation of reality.

Conclusion: Your Journey Into New Realities

U nderstanding parallel realities represents more than an intellectual journey—it offers a profound transformation in how we perceive ourselves and our place in the cosmos. Throughout this exploration, we've ventured from the solid ground of established physics into the fascinating realm of theoretical possibilities, discovering how the nature of reality might be far more expansive and intricate than we once imagined. This journey reveals that our everyday experience represents one thread in a boundless web of potential.

The concepts we've explored—quantum mechanics, parallel universes, multiple dimensions—aren't just abstract scientific theories. They provide a framework for understanding the true scope of existence and our role. When we grasp how reality might extend beyond our conventional understanding, we naturally expand our view of what's possible in our lives. This expanded perspective dissolves self-imposed limitations and opens new horizons of potential.

Exploring these ideas has equipped you with a unique lens

through which to view existence. Understanding parallel realities encourages us to question our assumptions, challenge our perceived limitations, and remain open to unexpected possibilities. This mindset naturally fosters creativity, resilience, and personal growth. Our approach to challenges and opportunities fundamentally shifts when we genuinely understand that reality might be infinite in its options.

The practical implications of this understanding extend into every aspect of life. Decision-making becomes less about finding the "right" choice and more about exploring different possibilities. Personal growth transforms from a linear journey to an expansive exploration of potential. Even our relationship with uncertainty changes—what once seemed threatening becomes an invitation to discover new aspects of reality.

These concepts also offer profound implications for human consciousness and potential. As we've discovered, the observer plays a crucial role in quantum mechanics, suggesting that consciousness itself might be fundamentally connected to the nature of reality. This understanding invites us to consider our role in shaping the reality we experience while maintaining humility about the vast mysteries that still await discovery.

The journey continues. Scientific understanding continues to evolve, offering new insights into the nature of reality. Emerging technologies provide new tools for exploration, while theoretical frameworks continue to expand our conception of what's possible. You're encouraged to remain curious and explore these ideas through scientific literature, philosophical discourse, and contemplation.

Consider this conclusion not as an endpoint but as a launching pad for further exploration. Our examined concepts offer endless opportunities for deeper understanding and personal growth. Whether through scientific study, philosophical inquiry, or practical application in your daily life, these ideas can continue to expand your perspective and enrich your experience of existence.

Remember that understanding parallel realities isn't about escaping our current experience but enriching it through a deeper appreciation of possibility. This knowledge empowers us to live more consciously, choose more deliberately, and approach life with incredible wonder and curiosity. It reminds us that each moment holds infinite potential, and our awareness of this potential enhances our ability to engage fully with life.

As you move forward, carry with you the understanding that reality is more mysterious and magnificent than we can fully comprehend. Let this knowledge inspire you to remain open to new possibilities, to question your assumptions, and to approach life with a sense of wonder. Your awareness of parallel realities can remind us that existence is more prosperous and full of potential than we typically imagine.

The power of an expanded mind lies not just in what it knows but in how it approaches the unknown. Let your understanding of parallel realities inspire you to remain curious, open, and engaged with the mysteries of existence. Your journey of discovery is ongoing, and each new understanding opens the door to further exploration and growth.

As you continue your personal journey, remember that you are both the observer and participant in this vast cosmic dance of possibility. Your choices, awareness, and engagement with life contribute to reality unfolding. This understanding brings both responsibility and opportunity—the responsibility to engage consciously with life and the chance to explore the full scope of human potential.

May this exploration of parallel realities serve as a foundation for continued growth, discovery, and wonder. The universe is more extraordinary than we can imagine, and your journey of understanding is just beginning. Embrace the mystery, remain open to possibility, and continue to explore the infinite potential landscape within and around you.

Recommended Resources

Books for Further Reading

B ooks for Further Reading

- "The Elegant Universe" by Brian Greene
- "The Hidden Reality: Parallel Universes and the Deep Laws of the Cosmos" by Brian Greene
- "The Fabric of Reality" by David Deutsch
- "Our Mathematical Universe" by Max Tegmark
- "Something Deeply Hidden: Quantum Worlds and the Emergence of Spacetime" by Sean Carroll
- "Parallel Worlds" by Michio Kaku
- "The Quantum Universe" by Brian Cox and Jeff Forshaw
- "Reality Is Not What It Seems" by Carlo Rovelli
- "The Order of Time" by Carlo Rovelli
- "The Big Picture" by Sean Carroll

Popular Science Books

- "A Brief History of Time" by Stephen Hawking
- "The Universe in Your Hand" by Christophe Galfard
- "Hyperspace" by Michio Kaku
- "The Science of Interstellar" by Kip Thorne

• "The Physics of the Impossible" by Michio Kaku

Movies That Explore Parallel Realities

• "Interstellar" (2014)
• "The Matrix" trilogy (1999-2003)
• "Everything Everywhere All at Once" (2022)
• "Doctor Strange" (2016)
• "Spider-Man: Into the Spider-Verse" (2018)
• "Coherence" (2013)
• "Sliding Doors" (1998)
• "Source Code" (2011)
• "Inception" (2010)
• "Mr. Nobody" (2009)
• "The One I Love" (2014)
• "Predestination" (2014)

Documentaries

• "What the Bleep Do We Know!?" (2004)
• "Through the Wormhole" (TV Series)
• "The Fabric of the Cosmos" (PBS Series)
• "Parallel Worlds, Parallel Lives" (BBC)
• "The Quantum Activist" (2009)
• "The Elegant Universe" (NOVA Series)
• "Into the Universe with Stephen Hawking" (Discovery Channel)
• "How the Universe Works" (Science Channel)
• "The Mystery of Matter: Search for the Elements" (PBS)
• "Beyond the Cosmos" (National Geographic)

TV Series

- "Fringe" (Science Fiction Series)
- "Dark" (Netflix Series)
- "Rick and Morty" (Animated Series)
- "The Man in the High Castle" (Based on Philip K. Dick's novel)
- "Parallels" (Science Fiction Series)
- "Sliders" (Classic Science Fiction Series)
- "Devs" (FX on Hulu)
- "The OA" (Netflix Series)
- "Counterpart" (Starz Series)
- "For All Mankind" (Apple TV+ Series)

Online Resources

- PBS Space Time (YouTube Channel)
- Quantum Gravity Research (YouTube Channel)
- Veritasium (YouTube Channel)
- World Science Festival (Website and YouTube)
- Sixty Symbols (YouTube Channel)
- arXiv.org (Scientific Paper Repository)
- Stanford Encyclopedia of Philosophy (Website)
- Closer to Truth (Website and PBS Series)

Call to Action

D ear Reader,

Thank you for joining me on this fascinating journey through parallel realities and infinite possibilities. Your willingness to explore these mind-expanding concepts speaks to your curiosity and desire for deeper understanding.

If you've enjoyed exploring these concepts, you might also be interested in my other book: "Simulation Theory for Beginners: Evaluating the Simulation Hypothesis and Its Virtual Reality Matrix." This companion volume delves into the intriguing possibility that our reality might be a sophisticated simulation, examining the evidence and implications of this mind-bending hypothesis. Together, these books provide a comprehensive exploration of the nature of reality and consciousness.

Share Your Experience

I'd love to hear about if either book has impacted your perspective or sparked new insights. Your feedback is invaluable not only to me as an author but also to other readers seeking knowledge in this area.

Leave a Review

Reviews play a crucial role in helping others discover these books. If you found value in these pages, please consider taking a few minutes to share your thoughts on Amazon. Your honest review, whether it's a few sentences or several paragraphs, can help:

- Guide other readers in their decision-making
- Provide valuable feedback for future editions
- Create meaningful dialogue about these fascinating concepts
- Help these books reach a wider audience

How to Leave a Review:

- Visit the book's page on Amazon
- Scroll down to the "Customer Reviews" section
- Click on "Write a customer review"
- Share your honest thoughts about the book
- Select a star rating that reflects your experience

Scan the QR Code to Leave a Review: Let Others Know What's Real

Your insights could unlock infinite possibilities for others. Scan to share your thoughts on 'The Beginner's Guide to Parallel Realities' and spark new conversations about the nature of our reality.

Share With Others

If you know someone who might benefit from these ideas, please share these books with them. The concepts of parallel realities and simulation theory become even more fascinating when explored and discussed with others.

Your Next Steps

As you close these pages, remember that this is just the beginning. I encourage you to:

- Reflect on the concepts we've explored
- Apply these ideas to your daily life
- Continue your research using the recommended resources
- Engage with others who share your interest in this field

III

Quantum Physics Unplugged: A Beginner's Guide to Understanding the Universe's Greatest Mysteries - Master the Basics Through Clear Language, Fun Examples, and Zero Complex Math

IV

Foundations of the Quantum Universe

IV

Foundations of the Quantum
Universe

Introduction

Y ou're wielding quantum physics every time you snap a photo with your smartphone. That crisp, vibrant selfie? It exists because of quantum mechanics working inside your phone's image sensor. Yet no one ever told you that, did they? This is just one example of how quantum physics quietly runs your daily life, even though most think it's locked away in research labs or theoretical papers.

I remember the first time I tried to learn quantum physics. I was staring at a textbook filled with intimidating equations and abstract concepts, feeling my confidence drain with each incomprehensible page. Maybe you've been there, too—trying to understand concepts like "quantum superposition" or "wave-particle duality" only to feel more confused than when you started. Perhaps you've wondered why something so fundamental to our modern world seems accessible only to those with advanced degrees in physics.

Here's the thing: quantum physics isn't just hiding in your smartphone camera. It's running your life in ways you'd never expect. When you use GPS to navigate to that new restaurant?

Quantum effects must be accounted for, or you'd end up miles off course. That LED screen you may be reading this on? Quantum mechanics makes those pixels light up. Are solar panels powering more and more homes? They work because of quantum physics. Photosynthesis, the process that gives us the oxygen we breathe, relies on quantum phenomena.

But if quantum physics is fundamental to our daily lives, why does it seem impossible to understand?

I've spent years watching brilliant minds get stuck at the same roadblocks. They pick up a quantum physics book, encounter a wall of mathematical equations, and put it down, feeling defeated. They watch online lectures that zoom past crucial concepts, leaving them more confused than enlightened. They read articles that either oversimplify everything into meaningless analogies or dive so deep into technical jargon that they might as well be reading ancient Sanskrit.

Think about the last time you tried to learn about quantum physics. It could be an article about quantum computers that caught your eye or a fascinating documentary about the nature of reality. You started reading or watching with excitement, ready to unlock the mysteries of the universe. But then came the avalanche of complex terminology: wave functions, quantum entanglement, superposition, quantum tunneling. Before you knew it, that initial excitement turned to frustration, and you wondered if this knowledge was beyond your reach.

This frustration isn't just about missing out on fascinating science – it's about feeling left behind in a world racing toward a quantum revolution. Quantum computers are emerging that could revolutionize medicine, crack current encryption methods, and transform artificial intelligence. Major companies are investing billions in quantum technology. Financial institutions are preparing for quantum-safe cryptography. Healthcare companies are exploring quantum computing for drug discovery. Tech giants are developing quantum sensors that could transform everything from medical imaging to navigation systems.

The quantum future isn't coming – it's already here, transforming industries in ways that would have seemed like science fiction just a decade ago. In hospitals worldwide, medical researchers are peering into the human body with unprecedented precision, using quantum sensors to spot diseases at the molecular level before they become life-threatening. Their work promises to revolutionize early detection and treatment, potentially saving millions of lives.

Meanwhile, in the digital realm, cybersecurity experts are racing against time. They know that quantum computers will soon be able to crack today's encryption methods, so they're developing new quantum-resistant security systems to protect our digital lives. It's like rebuilding the walls of a fortress while knowing that a powerful new weapon is being developed that could breach them.

Energy companies aren't sitting idle either. They're harnessing quantum computing to solve complex optimization problems that could make our power grids more efficient and help develop the next generation of batteries. These advancements could accelerate our transition to renewable energy and help combat climate change.

In the fields and farms that feed our world, agricultural scientists are deploying quantum sensors to monitor soil conditions and crop health with extraordinary precision. These tools allow farmers to understand their land at the molecular level, optimizing water usage and reducing the need for pesticides and fertilizers.

Perhaps most surprisingly, quantum technology is transforming the world of finance. Banks and investment firms are exploring quantum algorithms that could revolutionize risk analysis and fraud detection. These systems could spot patterns and anomalies that traditional computers might miss, making our financial system more efficient and secure.

Using the revolutionary S.P.A.C.E.T.I.M.E. Method, a framework that breaks down quantum physics into nine digestible concepts: Simplicity, Paradox, Ambiguity, Certainty limits, Entanglement, Transformation, Innovation, Medicine, and Expansion, this book will transform how you understand quantum physics. No complex math. No impenetrable jargon. Just clear explanations connected to the technology you use every day.

Think of this book as your quantum physics translator. Instead of drowning you in equations, I'll show you how quantum tunneling makes your computer's transistors work. Rather than abstract theories, you'll learn how quantum entanglement powers the next generation of ultra-secure communications. Each chapter builds your understanding through real-world examples and applications you can see and touch.

The journey through this book follows the same path our minds naturally take when learning something new: we start with what we know and gradually venture into the unknown. Think of it as exploring a fascinating new city. We'll begin on familiar streets, like the technology you use daily, before wandering down increasingly exotic alleyways of quantum understanding.

You might be surprised to learn that quantum physics is already your constant companion. Every time you check your location on your phone, quantum effects make that blue dot on your map more accurate. When you step outside and feel the sun's warmth on your face, you're experiencing the result of quantum processes that began deep in the sun's core. Even the plants in your garden rely on quantum tunneling to turn sunlight into energy through photosynthesis.

As we explore deeper, you'll discover why quantum computers aren't just faster versions of regular computers but fundamentally different machines that could revolutionize everything from drug discovery to climate modeling. We'll see how quantum technology can transform medicine, reshape the financial world, and create unhackable communication networks.

By the end of our journey, you'll find yourself looking at the world differently. The smartphone in your pocket will become more than just a device – you'll understand the quantum principles that make its components work. When you hear news about quantum computing breakthroughs, you'll grasp what they mean and why they matter. You'll be able to confidently join conversations about quantum technology, separating science fact from science fiction.

Most importantly, you'll develop an intuitive feel for quantum physics without drowning in complex mathematics. You'll see how quantum mechanics shapes our present and future, from the smallest microchip to the most significant questions about the nature of reality itself. And if you choose to dive deeper into more advanced quantum concepts, you'll have a solid foundation upon which to build.

This isn't just about adding new knowledge – it's about transforming how you see the world around you. The quantum realm might seem strange initially, but by the end of this book, you'll find it as familiar as your reflection.

The best part? You don't need any prior physics knowledge. If you can use a smartphone, you can understand quantum physics. This isn't about memorizing formulas – it's about seeing the quantum world that's been around you all along. It's about understanding the principles that will shape our technological future.

This book is for the curious minds who know there's more to quantum physics than Schrödinger's famous cat. It's for professionals who want to understand how quantum computing might transform their industry. It's for parents who wish to explain how the universe works to their children. It's for anyone who's ever looked at the mysteries of the quantum world and thought, "There must be a better way to learn this."

Most importantly, this book is for you if you've ever felt that quantum physics was beyond your reach. These concepts aren't reserved for physicists and mathematicians – they belong to everyone who wants to understand the fundamental workings of our universe and the technology that will shape our future.

Welcome to quantum physics unplugged. Let's decode the universe together, one concept at a time.

Simplicity: When Giants Stumble

I n 1897, physicist William Thomson (Lord Kelvin) made one of science's most infamous declarations: "There is nothing new to be discovered in physics now. All that remains is more and more precise measurement."

That same year, Ernest Rutherford discovered the electron—and suddenly, the entire foundation of physics began to crack. It was as if someone had confidently declared they'd mapped every street in New York City, only to discover an entire underground network of roads that didn't follow the usual traffic rules.

The discovery of the electron wasn't just another scientific breakthrough—it was the first glimpse into a hidden world that would forever change our understanding of reality. Imagine being a master chef who's perfected every recipe in your cookbook, only to discover an entirely new type of ingredient that doesn't follow any of your cooking rules. That's precisely what happened to physicists at the turn of the 20th century. The electron, this tiny particle that helps power your smartphone and makes lightning possible, refused to behave according to the established laws of physics. It was like finding a liquid that flows upward instead of down or a ball that bounces higher each time it hits the ground.

What is Quantum Physics?

If you've ever tried learning a new language, you know that moment when you realize you can't just translate word-for-word from English. Each language has its own rules, logic, and way of expressing ideas that might not even exist in your native tongue. Quantum physics is similar—it's the language of the incredibly small and comes with seemingly bizarre grammar.

Classical physics is like watching a football game from the stands. You can see the whole field, track the ball, and predict where players will move. But quantum physics? That's like trying to watch the same game through a keyhole. Suddenly, the rules that made perfect sense from above become confusing and unpredictable. You might see a player appear in one place. Seemingly teleport to another, or somehow be in two places simultaneously.

At its core, quantum physics is the science of the tiny atoms and the even tinier particles that make them up. But calling it "the science of the very small" is like calling the Internet "a bunch of connected computers." Technically accurate, but it misses the revolutionary impact it has on our world.

The quantum world isn't just some abstract realm confined to physics laboratories—it's woven into the fabric of your daily life in ways that would seem magical if you didn't know the science behind them. Take your morning commute, for instance. You rely on quantum effects when you check your phone's GPS to avoid traffic. Without accounting for these quantum influences, your location would drift by miles, sending you to the wrong exit or down the wrong street. Those satellites orbiting high above Earth need quantum mechanics to tell you precisely where you are.

Look closer at your phone's screen. Each pixel that lights up to form text, images, and videos works because of quantum mechanics' strangest feature: quantum tunneling. Particles in your phone's display do something that should be impossible— they pass through solid barriers as if walking through walls. It's not a glitch or a malfunction; it's a fundamental part of how our universe works at its most minor scales.

The natural world performs its quantum magic every day. As you walk past a tree or tend to your garden, you witness quantum mechanics. Those green leaves aren't just pretty— they're sophisticated quantum machines. Photosynthesis, the process that keeps plants alive and provides us with oxygen to breathe, achieves its remarkable efficiency by exploiting quantum effects. It's as if nature figured out quantum mechanics billions of years before we did.

Even your own body is quantum-powered. The DNA that makes you who you are occasionally undergoes mutations— changes that drive evolution itself. These mutations often occur because of quantum tunneling, the same ghostly phe-nomenon that makes your phone's screen work. Above us all, the sun continues to shine because quantum tunneling allows atomic nuclei to fuse, releasing the energy that lights and warms our world. Without quantum mechanics, the sun would go dark, and life as we know it would be impossible.

Quantum physics operates at a scale so small it's hard to imagine. If you magnify a hydrogen atom until its nucleus is the size of an orange, its electron would be a speck about 2.5 miles away. Yet, understanding this realm has led to technologies that define our modern world.

When Newton's Laws Weren't Enough

We all intuitively know Newton's laws. Throw a ball up; it comes down. Push a shopping cart; it moves forward. These laws work brilliantly for the world we can see. They're why engineers can build bridges that don't collapse and why NASA can land rovers on Mars.

Newton's laws elegantly explain the world around us. Watch a car roll to a stop at a red light, and you'll see his First Law of Motion in action—the car wants to keep moving forward until the brakes overcome its inertia. Every morning commute is a testament to Newton's genius.

When you watch a rocket launch on TV, you see Newton's Third Law performing its cosmic ballet. The massive thrust of flame shooting downward isn't just for show—it's the other half of the action-reaction pair that hurls spacecraft into the heavens. Each pound of thrust pushing down creates an equal force pushing the rocket up, carrying human dreams into space.

Step into a pool hall, and you'll find Newton's laws governing every collision and ricochet. The satisfying crack of a perfect break shot demonstrates the conservation of momentum as energy flows from the cue ball through the neat triangle of balls, sending them spinning across the green felt in predictable patterns.

Even something as thrilling as skydiving follows Newton's orderly rules. As a diver plummets through the air, gravity pulls it downward until air resistance builds to match it perfectly. At this equilibrium point, they reach terminal velocity—a perfect balance of forces Newton would have understood completely.

And far above us, planets dance around the sun in endless orbital loops, guided by the same law of universal gravitation that brings a tossed ball back to Earth. From the most miniature game of catch to the grandest cosmic ballet, Newton's laws seemed to explain everything—until we looked much closer.

But something strange happened when scientists tried to apply these same laws to the atomic world. Imagine you're playing pool. Newton's laws tell you exactly how the balls will move when collating. Now, shrink those pool balls to the size of atoms, and suddenly, nothing makes sense anymore. The balls might tunnel through each other, appear in two places at once, or somehow affect each other from across the table without touching.

Here are three dramatic examples where classical physics

completely broke down:

1. The Case of the Impossible Atom

According to Newton's laws, electrons orbiting an atom's nucleus should spiral inward and crash, like a satellite eventually falling to Earth. The math was straightforward: atoms shouldn't exist for more than a fraction of a second. Yet here you are, reading this book, made of atoms that have been stable for billions of years. Classical physics couldn't explain why you weren't imploding this very moment.

2. The Light That Shouldn't Work

Light hitting certain metals can knock electrons loose—this is called the photoelectric effect. Classical physics predicted that brighter light should give electrons more energy. Instead, the power depends on the color (frequency) of the light, not its brightness. It's like saying the temperature of your shower depends on the color of your shower head, not how far you turn the handle. This made no sense under classical physics.

3. The Hot Object Paradox

Classical physics insisted that hot objects emit radiation at all frequencies equally, releasing an infinite amount of energy. A simple hot cup of coffee should release enough power to destroy the universe. Spoiler alert: your morning coffee has yet to cause the apocalypse.

The Ultraviolet Catastrophe: When Physics Predicted the Impossible

The "hot object paradox" became known as the ultraviolet

catastrophe, and it was more than just an intriguing puzzle—it was a crisis that threatened to undermine all of physics. Imagine you're using a map app that's always been reliable. Still, one day, it tells you to drive your car off a cliff because that's technically the shortest route. Would you trust that app again? Would you question everything it had told you before?

This was the situation physicists faced. Their trusted theories, which worked perfectly for centuries, suddenly made impossible predictions. It was like having a calculator that correctly solved every math problem except those involving the number 7, where it would suddenly claim that 7 + 7 = infinity.

This was so important because everything in the universe emits radiation based on its temperature. The sun, your body, that cup of coffee on your desk—they're all constantly emitting energy. Classical physics had a formula for this and worked well for low frequencies (like radio waves). But when applied to higher frequencies (like ultraviolet light), the formula predicted that any warm object should emit infinite energy.

Please take a moment to consider the absurdity of what classical physics was telling us. That simple cup of coffee sitting on your desk—the one you're sipping while reading this book—should, according to the old laws of physics, be releasing enough energy to vaporize our entire planet. You're having breakfast next to a doomsday device every morning, yet here you are, peacefully stirring your coffee.

Flip on a light switch, and classical physics insists you've just triggered an apocalyptic event. That humble bulb in your lamp should flood the universe with infinite energy the instant it illuminates. Yet, your electricity bill remains finite, and your living room hasn't become a cosmic catastrophe.

The sun, that familiar yellow disk in the sky that's faithfully warmed our planet for billions of years, should have destroyed our entire solar system long before the first dinosaurs walked the Earth. According to classical physics, it should have unleashed an infinite inferno that would have made a supernova look like a sparkler.

Even more bizarre, the warmth of your body—that modest 98.6 degrees Fahrenheit—should be enough to trigger a catastrophe of cosmic proportions. Every living thing, from bacteria to blue whales, should be a walking astronomical disaster. Yet here we all are, our body heat doing nothing more dramatic than fogging windows on cold mornings.

Yet none of this happens. Your coffee typically gets cold, light bulbs work fine, and we're all still here. Something had to be fundamentally wrong with the classical understanding of physics.

Why Tiny Things Behave Differently

Think about throwing a tennis ball. Now imagine throwing a soap bubble. The tennis ball follows a nice, predictable path that Newton's laws describe perfectly. The soap bubble? It weaves and bobs, affected by the slightest air current. The smaller something is, the more different forces come into play.

Now, shrink to the size of an electron—about 1/100,000th the size of an atom, which is already incredibly tiny. At this scale, things get weird. Weird. An electron isn't like a small tennis ball or a soap bubble. It's more like nothing we experience in our everyday world.

Let me give you some mind-bending examples of how the quantum world differs from our everyday experience:

1. Location Uncertainty

In our world, your car is either in the garage or not. But an electron can be in multiple places simultaneously—imagine your car being in your garage and your neighbor's garage simultaneously until someone looks to check where it is.

2. Quantum Tunneling

It's as if you threw a ball at a brick wall, and sometimes the ball would pass right through to the other side, completely intact. This isn't science fiction—it's how solar fusion works and how some types of computer memory store information.

3. Action at a Distance

Imagine having two coins that always land on opposite sides, even on different continents. When you flip one, the other instantly responds. Einstein called this "spooky action at a distance" a natural phenomenon called quantum entanglement.

4. The Observer Effect

In our world, looking at something doesn't change it. But in the quantum world, the mere act of observation can alter the outcome of an experiment. It's like having a friend who always acts differently when they know you're watching.

The Birth of Quantum Theory: A Detective Story

The story of quantum physics reads like a scientific thriller, complete with unlikely heroes, dramatic revelations, and philosophical debates that continue to this day. Let's meet our cast of characters and see how they unveiled the quantum world.

The First Mystery: Black Body Radiation (1859-1900)

Gustav Kirchhoff started our story by studying how objects emit radiation when heated. It seemed simple enough—heat something, and it glows. But the mathematics produced nonsensical results, predicting infinite energy from finite objects.

Enter Max Planck, a conservative physicist who made a radical proposal out of desperation. Imagine if you could only withdraw money from your bank account in $10 increments—no amounts in between. Planck suggested energy worked the same way, coming only in discrete packets he called "quanta." He didn't believe this was true; he thought it was just a mathematical trick to make the equations work.

In 1905, a young patent clerk named Albert Einstein did something audacious with Planck's revolutionary idea. Where others saw a mathematical trick, Einstein saw a fundamental truth about light. He proposed that light wasn't just a wave flowing through space like ripples on a pond—it was also a stream of tiny particles, later called photons. This wasn't just splitting hairs; it solved mysteries that had puzzled scientists for years.

Scientists had long been baffled by how metals behaved when exposed to light. Shine a bright red light on certain metals, and nothing happens. But shine even a dim violet light, and electrons suddenly burst forth from the metal's surface. It made no sense—until Einstein explained that each photon carried a specific amount of energy determined by its color. Violet photons packed enough punch to knock electrons loose; red ones didn't, no matter how many you used. It was like discovering that throwing a single golf ball could break a window while throwing a thousand tennis balls wouldn't leave a scratch.

Eight years later, Niels Bohr took quantum ideas in an even more radical direction. He turned his attention to the atom—specifically, the puzzle of why atoms were stable. According to classical physics, electrons should quickly spiral into the nucleus, causing all matter to collapse. Yet here we are, solid as ever.

Bohr's insight was revolutionary: electrons couldn't orbit just anywhere. They were confined to specific energy levels, like athletes assigned to particular seats in a stadium, with no standing room allowed. When electrons jumped between these levels, they absorbed or emitted exact amounts of energy—explaining why heated elements always produced specific light colors. It was nature's light show, precisely choreographed by quantum rules.

Then came Werner Heisenberg in 1927 with perhaps the most profound insight. He discovered that the quantum world enforces a cosmic speed limit on knowledge. Want to know exactly where an electron is? OK, but you'll lose track of its speed. Want to measure its speed precisely? You'll have to give up knowing its exact location. This wasn't about having better tools or being more careful—it was a fundamental limit written into the fabric of reality.

This strange new quantum world needed interpretation, and in 1927, the greatest minds in physics gathered at the Solvay Conference to make sense of it all. What emerged was the Copenhagen Interpretation, a radical new way of thinking about reality. It suggested that quantum systems exist in multiple states simultaneously, like a cosmic game of possibility, until we measure them. The act of measurement forces reality to "choose" one definite outcome from all these possibilities.

Einstein, who had started this quantum revolution, couldn't accept its strange implications. "God does not play dice with the universe," he famously objected. With characteristic dry wit, Bohr reportedly replied, "Einstein, stop telling God what to do." Their debate would continue for decades, but the quantum die had been cast. Physics would never be the same.

All this talk of quantum mechanics might seem abstract—like

something trapped in the realm of theoretical physics—but look no further than your pocket to find it at work. That smartphone you carry isn't just a triumph of engineering; it's a quantum mechanical marvel. Every time you tap its screen, billions of transistors exploit quantum tunneling to process your commands. The bright, crisp display that lights up to show your messages? It works because we understand how electrons jump between precise energy levels in quantum systems. Even the solar panels charging your wireless earbuds harness Einstein's explanation of the photoelectric effect.

Step into any modern hospital, and you'll find quantum mechanics saving lives. MRI machines peer inside our bodies by choreographing the quantum dance of atomic nuclei. PET scanners use antimatter—positrons, the electron's mysterious twin—to spot diseases before they become visible through other means. Quantum dots light up cancer cells that would otherwise hide from view while precisely controlled laser beams, guided by quantum principles, perform surgeries with incredible precision.

Our connected world itself rests on quantum foundations. Every time you stream a movie or send a text, that information races through fiber optic cables as quantum particles of light. The GPS that guides you through city streets only works because it accounts for quantum effects on satellite clocks. These corrections are necessary for you to be allowed to your intended destination. And now, quantum cryptography promises to create truly unbreakable codes protected by the fundamental laws of physics.

But what we see today is just the beginning. We're standing on the threshold of a second quantum revolution. Quantum computers are beginning to solve problems that would take classical computers longer than the universe's age to crack. Quantum sensors promise to detect diseases at the molecular level, potentially spotting health issues years before symptoms appear. The quantum internet could make today's security concerns obsolete, creating communication networks protected by the strange rules of quantum entanglement.

In research labs worldwide, scientists are pushing the boundaries of what's possible. They're creating new quantum materials with properties that defy classical physics. They're developing sensitive quantum sensors that can detect a single atom out of place. They're building quantum computers that don't just process information differently—they harness the weirdness of the quantum world to perform computations that classical computers can't even approach.

And this is just the beginning. The quantum technologies we're developing today could transform every field, from medicine to climate science, financial modeling to artificial intelligence. We're learning not just to understand the quantum world but to shape it to our needs.

As we move forward, prepare yourself for something even more mind-bending. In the next chapter, we'll explore one of the most famous quantum experiments ever performed: the double-slit experiment. You'll see how single particles can exist in multiple places at once and how looking at them fundamentally changes their behavior. If you think what we've covered so far is strange, as Niels Bohr famously said, "If quantum mechanics hasn't profoundly shocked you, you haven't understood it yet." But don't worry—understanding why it's shocking is the first step to understanding how it works.

The quantum future isn't coming—it's already here, stranger and more wonderful than we could have imagined.

Paradox - Light's Double Life

W hat if I told you that light can simultaneously behave like a gentle ripple in a pond and a speeding bullet? As physicist Niels Bohr once said, "Anyone who is not shocked by quantum theory has not understood it." The story of light's double life is one of the strangest—and most important—discoveries in science, and it's about to blow your mind.

Imagine watching a superhero movie in which the main character can be as solid as steel and pass through walls like a ghost. Sounds impossible, right? Yet, this is exactly what light does in the real world. It's both a wave, spreading out like ripples in a pond, and a particle, behaving like tiny bullets. This isn't science fiction—it's the reality that powers everything from your smartphone's camera to the internet itself.

The Double-Slit Experiment: The Most Beautiful Experiment in Physics

Imagine being in London in 1801, watching a young scientist named Thomas Young set up what would become known as "the most beautiful experiment in physics." The setup was deceptively simple: a beam of light, two narrow slits cut into a card, and a screen to catch the light. But what this humble arrangement revealed would shake the foundations of physics and unveil one of nature's most profound mysteries.

To understand why this experiment was so revolutionary, picture throwing tennis balls at a wall with two narrow openings. Common sense tells us you'd find two neat piles of balls on the other side, each pile aligned with its respective slit. That's how particles should behave—they pick and stick to a path.

Now imagine dropping two pebbles into a calm pond. The ripples spread out, and something magical happens when they encounter a barrier with two gaps. The waves passing through these gaps spread out again, meeting and interacting. Where wave peaks meet, they create more significant peaks; where a peak meets a trough, they cancel each other out. This creates a distinctive pattern of alternating strong and weak ripples.

When Young shone his light through the slits, he saw something remarkable—bands of light and darkness, just like the interference pattern of water waves. It seemed to settle a centuries-old debate: light must be a wave, flowing through space like ripples on a cosmic pond.

But the story doesn't end there. Fast forward to the modern era, when scientists developed equipment sensitive enough to detect individual particles of light—photons. They decided to repeat Young's experiment, but with a twist: they would fire these photons one at a time, like shooting individual bullets through the slits. Indeed, they thought, each photon would pick one slit or the other, eventually building up two simple stripes of light.

What they saw instead defied all logic. Even firing single photons, one at a time, slowly built up the same interference pattern that Young had seen. It was as if each particle of light was somehow going through both slits simultaneously, interfering with itself. Imagine throwing a single tennis ball and somehow having it go through both holes at once— impossible in our everyday world but routine in the quantum realm.

The mystery deepened when scientists tried to catch these quantum acrobats in the act. They set up detectors to watch which slit each photon passed through. But the moment they did, the interference pattern vanished, replaced by the two simple stripes they had initially expected. It was like discovering you have a friend who performs impressive dance moves alone but instantly freezes into ordinary walking the moment you try to film them.

This isn't just some curious laboratory phenomenon. This same quantum wizardry—particles being waves, existing in multiple places at once, and changing behavior when observed—makes your digital camera work. It's what enables fiber optic cables to carry vast data globally. The "impossible" has become not just possible but practical.

Even Einstein, who had revolutionized physics with his theories of relativity, struggled to accept these implications.

But nature has little regard for what we find reasonable. In the quantum world, the impossible isn't just possible—it's mandatory.

Einstein and the Photoelectric Effect: Light Gets Particle-ar

While Young's dancing light waves captivated the scientific world, a different experiment quietly challenged everything physicists thought they knew about light. It started with a simple observation: when light strikes metal, it can knock electrons loose from their atomic bonds. Scientists called this the photoelectric effect, though it might as well have been called "light's particle punch."

According to the physics of the time, this effect should have been straightforward. Shine any light color on metal long enough; eventually, you'd provide enough energy to free some electrons. It was like pushing a child on a swing—even gentle pushes, given enough time, should eventually get the swing moving.

But nature had other ideas. When scientists experimented, they found something bizarre. A dim ultraviolet light would instantly send electrons flying from the metal surface. In contrast, even the brightest red light wouldn't budge a single electron, no matter how long it shone. Making the light brighter only produced more freed electrons, not faster-moving ones. It was as if the universe was playing by rules no one understood.

Enter Albert Einstein, who proposed something revolutionary: What if light wasn't just a wave but also came in tiny packets of energy, like miniature bullets? These packets—which we now call photons—would carry different amounts of energy depending on their color. Ultraviolet photons packed a powerful punch, while red photons were comparative weaklings.

Suddenly, the photoelectric effect made perfect sense. Each ultraviolet photon carried enough energy to knock an electron free immediately like a well-aimed golf ball breaking a piggy bank. Red photons, no matter how many you used, were like throwing ping pong balls at the bank—they didn't have enough individual energy to break anything loose. Brighter light meant more photons, like throwing more golf balls, but each electron was still only hit by one photon at a time.

This discovery, which would earn Einstein the Nobel Prize, revealed light as nature's ultimate shape-shifter. Like Mystique from X-Men morphing between different forms, light could act as either a wave or particle, depending on the situation. Light spreads out like a wave when traveling through space, exploring all possible paths. But when it interacts with matter—striking a metal surface or hitting your digital camera's sensor—it behaves like a particle, delivering its energy in discrete chunks.

This dual nature isn't a flaw or a contradiction—it's fundamental to how our universe works at its deepest level. We see it in action every time we use the internet. In fiber optic cables, light travels as waves, spreading through glass fibers with minimal loss. But when that light reaches its destination, photon detectors catch individual particles of light, converting them into the electrical signals that become your streaming videos, social media posts, and email.

Light, it turns out, isn't confused at all—we were the confused ones, trying to force it into our limited categories of "wave" or "particle." Nature refused to be so easily labeled, teaching us that reality is far richer and stranger than we had imagined.

Tony Stark Would Be Impressed: Modern Technologies That Use Light's Dual Nature

Remember the Arc Reactor from Iron Man? While Tony Stark's power source remains in science fiction, the fundamental technologies that harness light's dual nature are equally impressive. They're transforming our world in ways that make even Iron Man envious.

Take solar panels, for instance. These sleek energy collectors work because we understand both faces of light. Photons that strike the panel's surface behave like particles, knocking electrons loose through the photoelectric effect. Yet engineers design these panels by treating sunlight as a wave, carefully tuning them to capture specific wavelengths for maximum efficiency. This marriage of light's dual nature has made solar power a clean energy source increasingly viable.

Your digital camera tells a similar story. When you take a photo, light waves focus through your camera's lens system, bending and converging like ripples in a pond. But when this light hits the sensor, individual photons strike like tiny bullets, triggering precise electrical signals in each pixel. Your selfies exist because we've learned to manipulate both aspects of light's personality.

Lasers represent the most dramatic mastery of light's dual nature. By synchronizing light waves and controlling the flow of photons, we create intense beams that can cut through steel, perform delicate eye surgery, or read the microscopic pits in a Blu-ray disc. The same technology guides precision manufacturing, measures distances in construction, and enables fiber optic networks that form the backbone of the internet.

Speaking of displays, your phone's screen is a quantum light show. When electrons in the LED display jump between different energy states, they release precisely controlled photons of specific colors. These quantum leaps create every image you see, from your favorite social media feed to streaming videos.

The Quantum Revolution in Your Pocket

Your smartphone isn't just a communications device—it's a masterpiece of quantum engineering. The camera system alone brilliantly demonstrates this: light waves focus through precisely engineered lenses before individual photons trigger quantum effects in the sensor. The image processing software even accounts for quantum noise to produce clearer pictures.

The screen technology is equally sophisticated. Quantum dots—tiny semiconductor particles—create vivid colors by controlling how electrons jump between energy levels. LED backlights use similar quantum transitions to illuminate the display, while sophisticated light guides ensure even brightness across the screen.

In newer phones, even facial recognition relies on quantum effects. Infrared photons map your face's contours, quantum sensors detect the reflected light, and neural networks process this quantum information to verify your identity. Even your GPS location depends on quantum precision—satellite atomic clocks must account for both quantum and relativistic effects to give you accurate directions to the nearest coffee shop.

Beyond Consumer Tech: Medical and Scientific Applications

Light's quantum nature is improving our gadgets and revolutionizing medicine and scientific research. In hospitals, MRI machines peer inside our bodies by manipulating the quantum properties of atoms. PET scanners track quantum interactions to reveal metabolic processes. At the same time, X-ray machines precisely control photon energy to create detailed images while minimizing radiation exposure.

In research laboratories, quantum microscopes achieve the seemingly impossible, revealing individual atoms. Laser spectroscopy analyzes materials by examining their interaction with light at the quantum level. Perhaps most impressively, quantum sensors can now detect gravitational waves—ripples in spacetime—by measuring quantum effects in laser beams.

The Sun's Quantum Dance

Even our sun, that ancient nuclear furnace in the sky, depends on quantum effects for its existence. Nuclear fusion in the sun's core is only possible because of quantum tunneling—particles doing something classical physics says is impossible. Once created, photons begin an epic journey, taking thousands of years to reach the surface as they bounce through the sun's interior. When they finally emerge, light's wave-particle nature helps create everything from the sun's visible surface to our blue sky.

The future looks even brighter as we develop new quantum technologies. Quantum computers promise to solve problems beyond classical computers' reach. Quantum encryption could secure communications. Ultra-precise quantum measurements might reveal new phenomena we haven't even imagined yet.

Deeper Into the Quantum Realm

If light's ability to be both a wave and a particle seems strange, it's only the beginning. In the next chapter, we'll dive into one of quantum physics' most profound principles: uncertainty. You'll discover why particles don't have defined locations until we measure them and how this fundamental fuzziness of reality leads to remarkable technologies.

Remember: Your phone works because of quantum physics. So do your TV, LED lights, and car's GPS. The sun shines because of quantum tunneling. MRI machines see inside you using quantum effects. Solar panels make electricity through the photoelectric effect. Computer chips process this text using quantum mechanics.

That's fantastic about quantum physics—it's not just equations on a chalkboard. Science makes our modern world possible, and it's working its quantum magic in your pocket right now.

As we venture deeper into quantum physics, keep in mind that these "impossible" behaviors of light were just the beginning. Nature has many more surprises in store, and understanding them has given us technological superpowers that would make even Tony Stark jealous.

Ambiguity - The Probability Cloud

"Everything we call real is made of things that cannot be regarded as real." - Niels Bohr.

Let that sink in for a moment. One of the greatest physicists ever told us that reality itself isn't exactly... real. At least, not in the way we usually think about it. If that sounds mind-bending, you're in good company. Even Einstein struggled with this idea.

Imagine you're watching a coin spinning on a table. Before it lands, is it heads or tails? While it's spinning, it's somehow both and neither simultaneously. Now, shrink that coin down to the size of an atom, and you're beginning to grasp one of quantum physics' most profound ideas: at its most miniature scale, nature deals in possibilities rather than certainties.

Wave Functions: The Crystal Ball of Quantum Physics

Let's start with something simple: Where is an electron in an atom? You might remember pictures of electrons orbiting the nucleus like tiny planets around a sun in your school days. It's a nice image, but it needs to be corrected.

In quantum physics, we can't say exactly where an electron is. Instead, we use a wave function—a mathematical description of where it might be. Think of it like your weather app predicting tomorrow's forecast. When it shows a 70% chance of rain, it's not saying it will rain; it's just that it's likely. A wave function works similarly, telling us the probability of finding an electron in different locations.

Consider your local mall on a Saturday afternoon. Suppose someone asks where your teenage child is. In that case, there's a high probability they're at the food court, a decent chance they're at the gaming store, and a slim possibility they're doing homework in the library. A wave function describes an electron's location similarly—but with one crucial difference: until you look (measure its position), the electron is somehow in all those places simultaneously.

Welcome to the Probability Cloud

Forget everything you learned about electrons orbiting atoms like planets around the sun. The reality is far more fascinating. Instead of following fixed paths, electrons exist in "probability clouds" or orbitals. These aren't actual clouds you can see but mathematical descriptions of where an electron will likely be found at any given moment.

Think of it like predicting where a bee might be in a garden. You can't know its exact location, but it will likely be near flowers—the more attractive the flower, the higher the probability of finding the bee there. Until you spot the bee, it's as if it's visiting all the flowers at once—at least in the quantum world.

This isn't just theoretical musing—it's how atoms work. The shapes of these probability clouds determine how atoms bond, which determines the properties of every material around us. This understanding has revolutionized technology, leading to the development of semiconductors that power your phone, LED lights that illuminate your home, and solar panels that convert sunlight into electricity.

Schrödinger's Cat: The Most Famous Feline in Physics

Now we come to physics' most famous thought experiment: Schrödinger's Cat. Erwin Schrödinger devised this scenario to show how weird quantum mechanics becomes when applied to everyday objects.

Picture this: a cat is placed in a sealed box with a clever but diabolical device. Inside this box is a vial of poison, a radioactive atom, and a mechanism that will break the vial if the atom decays. The radioactive atom has a 50% chance of decaying within an hour. If it decays, the poison is released, and the cat dies. If it doesn't decay, the cat lives.

According to quantum mechanics, something extraordinary happens before we open the box. The radioactive atom exists in a superposition of both decayed and not decayed states. Consequently, the poison is both released and not released, and the cat is—impossibly—both alive and dead at the same time. Only when we open the box and observe the system does this superposition "collapse" into one definite outcome.

187

Of course, Schrödinger wasn't suggesting we put cats in boxes. He created this thought experiment to highlight the absurdity of applying quantum mechanical principles to everyday objects. The cat paradox illustrates a fundamental question in quantum physics: Where exactly does the quantum world end and our classical world begin? Why don't we see larger objects like cats in multiple states simultaneously?

Why Quantum Strangeness Makes Perfect Sense (Sort Of)

Despite its apparent strangeness, quantum mechanics is the most successful scientific theory ever developed. Every prediction it's made has been confirmed by experiments, and every technology based on its principles works exactly as expected. It's like having a foreign phrasebook that sounds nonsensical when read aloud but somehow gives you perfect directions every time you use it.

Take your smartphone's processor, for instance. Its billions of transistors work because we understand how electrons move through probability clouds. MRI machines create detailed images of your body by manipulating the quantum properties of atoms. The GPS in your car knows your location precisely because it accounts for quantum effects in satellite signals.

The fact that quantum mechanics seems strange to us isn't a flaw in the theory—it's a limitation of our human experience. We evolved to survive in a world of large objects where quantum effects average out. Expecting quantum physics to match our everyday intuition is like expecting a fish to understand the concept of drought.

Quantum Superposition: More Than Just a Parlor Trick

The idea that particles can exist in multiple states simultaneously might seem like a curious oddity. Still, it's becoming the foundation of a technological revolution. Quantum computers, for instance, harness this very principle to perform calculations that would be impossible for traditional computers.

While a classical computer bit must be either 0 or 1, a quantum bit (qubit) can simultaneously be in a superposition of both states. You could complete a multiple-choice test by choosing all answers simultaneously, then get the right one when it's time to turn in your paper. This isn't science fiction—companies like IBM, Google, and Microsoft are already building quantum computers based on these principles.

The Cloud Computing You Didn't Know About

The electron probability clouds we discussed earlier don't just help us understand atoms—they're essential to modern chemistry and materials science. When scientists develop new materials, they use quantum mechanics to predict how electron clouds will interact. This has led to breakthroughs in:

- Solar cell efficiency
- Battery technology
- Drug development

- Clean energy catalysts
- Advanced electronics

Every time you use a touchscreen, you interact with materials designed using our understanding of electron probability clouds. This same quantum foundation will emerge from the next generation of flexible electronics, quantum sensors, and clean energy technologies.

Living in a Quantum World

Despite its microscopic scale, quantum probability affects our daily lives in countless ways. The sun only shines because quantum tunneling allows nuclear fusion to occur. Plant photosynthesis uses quantum coherence to convert sunlight into energy more efficiently than any human-made solar cell. Even our sense of smell might rely on quantum effects.

Understanding quantum probability has deepened our knowledge of nature and given us technological superpowers our grandparents couldn't have imagined. Every digital photo you take, every LED light you switch on, and every semiconductor device you use works because we've learned to harness the quantum realm.

The Ultimate Speed Limit

We've seen how particles exist in multiple states until observed, how electrons live in probability clouds rather than fixed orbits, and how these strange properties enable modern technology. However, we must discuss another fundamental limit in quantum physics that Einstein himself struggled to accept.

In the next chapter, we'll explore Heisenberg's Uncertainty Principle, which limits what we know about quantum systems. We'll discover why we can never simultaneously know where a particle is and where it's going with perfect precision. This isn't a limitation of our measuring devices—it's a fundamental reality feature.

Remember, the quantum world is still being determined because we lack precise measurements. It's uncertain because uncertainty is woven into the fabric of reality. As we'll see, this inherent fuzziness of nature isn't a bug—it's a feature that makes our universe possible.

The journey into quantum ambiguity might seem strange, but it's leading us to a more profound truth about nature. As Niels Bohr said, "If quantum mechanics hasn't profoundly shocked you, you haven't understood it yet." But understanding this shock is the first step toward appreciating our remarkable world—a world where probability clouds dance in every atom of your body, and uncertainty becomes a certainty

V

Mind-Bending Quantum Concepts

Certainty Limits - The Edge of Knowledge

"Not only is the Universe stranger than we think, it is stranger than we can think." - Werner Heisenberg.

These words weren't just philosophical musing—they came from the physicist who discovered one of nature's most profound secrets: there are fundamental limits to what we can know about the universe. Not because our instruments aren't precise enough or because we're not smart enough, but because uncertainty is woven into the very fabric of reality.

When Certainty Crumbled

In 1925, a young Werner Heisenberg retreated to a windswept island in the North Sea, seeking relief from his hay fever. What he discovered there would shake the foundations of physics more profoundly than anything since Newton. Working late into the night, fueled by coffee and confusion, Heisenberg realized something revolutionary: the more precisely you measure specific properties of a particle, the less precisely you can know others.

This wasn't just another scientific discovery—it was a fundamental limit on knowledge itself. Einstein, who had revolutionized physics with relativity, found this idea deeply troubling. "God does not play dice," he famously objected. But Heisenberg's uncertainty principle wasn't about God or dice—it was about the nature of reality itself.

The Pool Hall That Changed Physics

Imagine you're playing pool. The green felt stretches before you, dotted with perfectly spherical balls. In this familiar world, everything follows common sense rules. You can know exactly where each ball is and how fast it moves. You can calculate perfect shots because classical physics gives you all the necessary information.

Now imagine shrinking to the size of an atom, cue stick in hand, trying to play pool with electrons instead of balls. Everything you know about pools breaks down. Try to measure where your electron "ball" is, and you lose track of its speed. Focus on its speed, and its location becomes a blur. This isn't because the electron pool is tough—it's because the universe prevents you from knowing both properties simultaneously.

Think about a game of pool for a moment. It's a perfect demonstration of classical physics in action. You can watch the white cue ball roll across the green felt, tracking its position at every instant. You can measure how fast it's moving, calculate the perfect angle for your shot, and predict where each ball will end up after the collision. Every aspect of the game

196

follows neat, predictable rules. Advanced players can even plan multiple shots, knowing that the physics of billiards will faithfully execute their strategic vision.

This predictability is what makes Pool such a satisfying game. Whether you're calling your shot on the eight ball or setting up a complex bank shot, you're relying on the clockwork nature of classical physics. Each click of the ball against the ball, each rolling trajectory, and each gentle spin follows laws so reliable you could set your watch by them. It's a world where uncertainty has no place, where cause and effect play out like a well-rehearsed dance.

But in the quantum world, we can never have complete information. The more precisely we measure position, the more uncertain momentum becomes, and vice versa. It's not a matter of building better-measuring tools or being more careful—it's a fundamental limit written into the laws of physics.

The Camera That Couldn't

To understand why this happens, look at something familiar: photographing a moving object. When you try to capture a runner sprinting past, you face a choice:

You can use a fast shutter speed to freeze the motion, getting a crystal-clear image of where the runner is. But that sharp image tells you nothing about how fast they move—they could be running or standing still. The position is specific, but the velocity is lost.

Or you can use a slow shutter speed, creating that artistic blur that shows motion. Now, you can tell precisely how fast they're moving by measuring the length of the blur. Still, you need precise information about their location at any specific moment. The velocity is captured, but the position becomes uncertain.

This photography analogy isn't perfect, but it hints at something profound about the quantum world. In quantum mechanics, this trade-off isn't just a limitation of our cameras—it's a fundamental property of reality. The universe works like a cosmic camera that can never get a perfect position shot and a perfect motion shot of its minor components.

Why We Can't Know Everything

The reason for this uncertainty goes deeper than just measurement problems. When scientists try to measure an electron's position, they must bounce light off it. But light isn't just illumination—it carries energy and momentum. The very act of measurement changes what's being measured.

Imagine figuring out where a soap bubble is by throwing ping-pong balls at it. Sure, when a ping-pong ball hits the bubble, you know where the bubble was—but you've also popped it! The quantum world is similarly delicate. Every measurement interaction changes the system being measured.

This is more than a technical problem we might solve with better instruments someday. Heisenberg proved that this uncertainty is built into the mathematical structure of quantum mechanics. It's not that we're not clever enough to measure both properties—it's that these properties cannot simultaneously have definite values.

When Einstein Was Wrong

This revelation sparked one of the most famous scientific debates in history. Einstein, troubled by this inherent uncertainty, spent years trying to prove Heisenberg wrong. He devised increasingly clever thought experiments, attempting to show ways to measure both position and momentum precisely.

However, Heisenberg and his colleagues found flaws in Einstein's reasoning each time. The uncertainty principle held firm. Eventually, Einstein's attempts to disprove quantum uncertainty led to even deeper insights into the quantum nature of reality.

The Universe's Fuzzy Edge

This principle is revolutionary because it reveals a fundamental fuzziness in nature. Before quantum mechanics, scientists believed particles were like tiny billiard balls—small but with definite positions and velocities at all times. The uncertainty principle showed this view was wrong.

Instead of definite properties, particles exist in states of possibility until measured. It's not that the electron is in a specific place. Still, we don't know where—it doesn't have a definite position until we measure it. The same goes for its momentum.

Nature's Impossible Trick: Quantum Tunneling

When you think quantum mechanics couldn't get any stranger, consider this: particles regularly do something that should be impossible—they pass through solid barriers. Physicists call this quantum tunneling, and it's not just a curious oddity. Without it, the sun wouldn't shine, and life wouldn't exist.

Imagine rolling a ball up a hill. If it doesn't have enough energy to reach the top, it rolls back down—every time. That's classical physics. But in the quantum world, something extraordinary happens. Sometimes, particles reach the other side of the "hill" even when they don't have enough energy to climb it. It's as if you threw a tennis ball at a brick wall, and occasionally, it would pass right through, emerging unchanged on the other side.

This isn't magic—it's pure quantum mechanics. Remember the uncertainty principle we just discussed? Because a particle's position is uncertain, there's always a tiny chance it will appear on the other side of a barrier. The particle didn't break through or climb over—it "tunneled" through a region it couldn't access.

The Sun's Quantum Secret

Here's something remarkable: every sunbeam you've ever felt exists because of quantum tunneling. At the sun's core, hydrogen nuclei must get extremely close to fuse into helium.

The problem is that these nuclei are positively charged, repelling each other violently. Classical physics says they could never get close enough to fuse—the energy barrier is too high.

But quantum tunneling saves the day. Because of uncertainty and tunneling, there's a tiny chance for these nuclei to appear close enough to fuse. This chance is so tiny that any particular pair of nuclei might wait billions of years to fuse. But with the astronomical number of nuclei in the sun, enough tunneling events happen every second to power all life on Earth.

Think about that moment: the light warming your skin now exists because particles did something that classical physics says is impossible.

Tunneling in Your Pocket

You don't need to look to the stars to find quantum tunneling—it's happening right now in your pocket. Every smartphone contains billions of transistors, many of which rely on quantum tunneling to function. Flash memory, which stores your photos and messages, works because electrons tunnel through fragile insulating barriers.

Modern electronics increasingly depend on this quantum "impossibility":

Scanning Tunneling Microscopes (STM) use quantum tunneling to create images of individual atoms. When the microscope's tip gets close enough to a surface, electrons tunnel between them; scientists can map surfaces with atomic precision by measuring these tunneling currents.

Your phone's processor contains tiny transistors, so quantum tunneling becomes both a challenge and a tool. Engineers must carefully control where tunneling happens and where it doesn't, turning quantum weirdness into practical computing power.

When Barriers Aren't Barriers

The implications of quantum tunneling go far beyond technology. It forces us to reconsider what we mean by a "barrier." In the classical world, barriers are absolute—either you have enough energy to overcome them, or you don't. But quantum mechanics reveals that barriers are more like probability filters. Usually, they keep things out, but there's always a chance of tunneling through.

This probabilistic nature of barriers appears throughout nature:

In biology, enzymes use quantum tunneling to speed up chemical reactions that would be impossibly slow otherwise. When hydrogen atoms need to transfer between molecules, they often tunnel through energy barriers rather than climb over them.

Some types of radioactive decay only happen because of quantum tunneling. The particles inside unstable nuclei occasionally tunnel out, leading to the steady tick of radioactive decay that helps us date ancient artifacts and power nuclear medicine.

Even the simple act of water molecules forming and breaking bonds involves quantum tunneling of hydrogen atoms. The quantum world is constantly at work in the most mundane aspects of our lives.

Living with Uncertainty

Together, the uncertainty principle and quantum tunneling reveal something profound about reality. The universe isn't the precise clockwork imagined by classical physics. Instead, it's built on a foundation of uncertainty and possibility.

This doesn't mean science is uncertain—quite the opposite. Quantum mechanics makes the most precise predictions of any scientific theory ever developed. We can calculate quantum effects with extraordinary accuracy. What's uncertain isn't our knowledge of the laws of physics but rather the inherent behavior of the particles themselves.

Modern technology increasingly embraces this quantum uncertainty. Quantum cryptography uses the uncertainty principle to create unbreakable codes. Quantum computers manipulate uncertain states to solve problems classical computers can't tackle. The future of technology may depend on learning to work with uncertainty rather than fighting against it.

The Edge of Knowledge

As we push the boundaries of science and technology, we keep running into these quantum limits. They remind us that the universe is stranger and more mysterious than our everyday experience suggests. But these limits haven't stopped progress—they've opened new doors.

Every time you use your phone, watch an LED screen or feel the sun's warmth, you're experiencing the effects of quantum uncertainty and tunneling. The same principles that limit our knowledge also enable our modern world.

When Particles Dance Together

We've seen how nature limits what we know about one particle. But something even stranger happens when we look at two particles together. They can share information instantly, no matter how far apart they are. Einstein called it "spooky." We call it quantum entanglement.

In the next chapter, we'll explore this mysterious connection between particles. This phenomenon challenged Einstein's view of reality and promised to revolutionize computing and communication. Get ready to discover why the quantum world isn't just uncertain—it's "spooky" too.

Entanglement - The Invisible Thread

"What happens to one particle can impact what we expect the second one to do, even if those particles are nowhere near each other." - David Kaiser.

Imagine having a pair of magical dice. You give one to your friend in Tokyo and keep one for yourself in New York. Every time you roll your die, your friend's die shows precisely the same number—instantly, with no communication between you. No radio signals, internet connection, or hidden transmitters—just two dice sharing an invisible bond across the planet. Impossible? Welcome to the world of quantum entanglement, perhaps the strangest feature of quantum physics and what Einstein famously called "spooky action at a distance."

The Twin Particle Mystery

In the quantum realm, particles can become "entangled" so that the fate of one is inextricably linked to the other, no matter how far apart they are. It's as if they share a secret connection that transcends space itself. When you measure one particle, you instantly know something about its distant partner.

Think of it as having identical twins who always wear opposite-colored socks without coordinating. If you see one twin wearing black socks in London, you instantly know their sibling in Sydney is wearing white socks. But even this analogy needs to capture the true strangeness of quantum entanglement. The socks were already black or white with the twins before you looked. In quantum entanglement, observing one particle determines the state of both particles at the exact moment.

Let's try another analogy. Imagine you have two coins sealed in separate envelopes. In the classical world, each coin is either a head or tail before you open the envelopes. But with quantum entanglement, it's as if the coins exist in a shared state of possibility until you open one envelope. At that instant, both coins snap into definite, correlated states, no matter how far apart the envelopes are.

Einstein's Spooky Action

When Einstein first encountered this idea, he was distraught. Here was the man who revolutionized our understanding of space and time with relativity, and he couldn't accept what quantum mechanics suggested about reality. His particular theory of relativity had shown that nothing, not even information, could travel faster than light. Yet here was quantum mechanics, suggesting that entangled particles could influence each other instantaneously across any distance.

Einstein believed the universe should make sense. Their immediate surroundings should only affect things like ripples spreading across a pond. The idea that measuring a particle in New York could instantly affect another particle on the moon seemed to violate everything he believed about how the universe should work.

In 1935, Einstein and his colleagues Boris Podolsky and Nathan Rosen published a paper that would become famous as the EPR paradox. They proposed a thought experiment that showed that quantum mechanics must be incomplete. Their argument was subtle but powerful: either quantum mechanics was wrong, or information could travel faster than light. Since Einstein had proven the latter was impossible, quantum mechanics had to be missing something.

Einstein suggested there must be "hidden variables"—some more profound layer of reality that would explain these impossible connections. In his view, quantum mechanics was like a weather forecast that could only predict probabilities. Just as a 70% chance of rain doesn't mean it's simultaneously raining and not raining, Einstein believed particles must have definite properties all along, even if we can't measure them all at once.

Bell's Brilliant Breakthrough

For nearly thirty years, the debate remained philosophical. How could you prove whether Einstein was right or wrong about something fundamental to reality? The answer would come from an unlikely source: an Irish physicist named John

Bell, who worked in his spare time while developing particle accelerators at CERN.

In 1964, Bell found a way to settle the debate experimentally. He derived a mathematical inequality that could distinguish between Einstein's hidden variables and the predictions of quantum mechanics. If Einstein was right, certain correlations between entangled particles couldn't exceed a specific value. If quantum mechanics was correct, that limit could be violated.

The genius of Bell's work was that it turned a philosophical question into a testable prediction. It was like finding a way to scientifically test whether love exists by measuring its effects on the physical world. Bell showed that Einstein's common-sense view of reality made subtly different predictions from quantum mechanics.

To understand Bell's test, imagine our magical dice again. If the dice were using hidden variables—some pre-arranged pattern or synchronization—there would be limits to how perfectly they could correlate under different conditions. However, quantum mechanics predicted stronger correlations than should be possible with any hidden mechanism. It was as if the dice weren't just synchronized but shared a single existence despite being continents apart.

The scientific community realized Bell's work's profound implications. Here was a way to test whether quantum mechanics was complete, as its defenders claimed, or whether Einstein was right about hidden variables. If experiments could be designed carefully enough, nature would tell us which view was correct.

Proving Einstein Wrong: The Experiments

In 1972, physicist Alain Aspect and his team at the University of Paris set up an experiment that would shake the foundations of physics. The challenge was enormous: they needed to create pairs of entangled particles, send them in opposite directions, and measure their properties before any possible signal could pass between them. It was like trying to prove that twins could share thoughts while ensuring they had no way to communicate.

The experimental setup was ingenious. Using calcium atoms excited by lasers, they created pairs of entangled photons—particles of light that shared a quantum connection. These photons were sent racing away from each other through fiber optic cables. The team used unique crystals at precisely timed intervals to measure the photons' polarization—the direction in which their light waves oscillated.

If Einstein had been right about hidden variables, these measurements would have shown a limited correlation. Think of it like flipping two somehow synchronized coins before being separated. No matter how clever the synchronization, there's a mathematical limit to how often the coins can match up under different flipping conditions. Bell's inequality spelled out precisely what this limit should be.

But the photons violated this limit spectacularly. Their properties matched more often than any hidden variable theory could explain. It was as if the photons were consulting an invisible playbook transcending space and time.

Scientists, being scientists, were only partially satisfied. What if there were loopholes in the experiment? They identified several possibilities:

The Locality Loophole: What if some subtle signal could still pass between the particles? Teams started separating their detectors by increasing distances. In 2015, researchers at Delft University conducted tests with particles 1.3 kilometers apart. The spooky action persisted.

The Detection Loophole: What if the measuring devices were somehow selecting only the particles that would prove quantum mechanics right? Scientists developed more efficient detectors and statistical methods to ensure they weren't fooling themselves.

The Freedom-of-Choice Loophole: What if the universe somehow conspired to predict which measurements would be made? Researchers used random number generators based on distant stars to decide how to measure each particle, ensuring even the light from those stars couldn't have reached the other detector in time to influence it.

These loopholes were closed one by one. In 2018, an international team used light from distant quasars—ancient galaxies billions of light-years away—to determine their measurements. Even then, the quantum correlations held firm. Einstein's common-sense view of reality had been experimentally disproven not just once but hundreds of times under increasingly stringent conditions.

Creating Quantum Connections

How do we create entangled particles? It's not as mysterious as it might sound, though it does require precise control over quantum systems. Scientists have developed several methods, each with its advantages:

Spontaneous Parametric Down-Conversion (SPDC): This tongue-twister process uses unique crystals to split single photons into pairs of entangled photons with lower energy. Imagine breaking a playing card perfectly down the middle— the two halves will always match perfectly. The crystal does the same thing to light, creating photon pairs with a quantum connection.

Atomic Cascades: When excited to high energy states, some atoms emit pairs of photons as they return to their ground state. These photon pairs are naturally entangled, like twin children born simultaneously.

Quantum Dots: These tiny semiconductor structures can trap electrons and force them to emit entangled photons. It's like having a quantum light bulb that produces entangled particles on demand.

Recent breakthroughs have pushed the boundaries of what's possible. In 2023, Princeton physicists achieved something remarkable—entangled individual molecules for the first time.

Using laser cooling to bring molecules nearly to absolute zero, they employed "optical tweezers" (focused laser beams that can trap and move tiny objects) to position molecules with incredible precision. Then, using carefully timed microwave pulses, they created quantum connections between these molecules.

Meanwhile, at CERN's Large Hadron Collider, scientists observed quantum entanglement between top quarks—the heaviest known elementary particles—and their antimatter counterparts. This showed that quantum effects don't just apply to tiny particles; they can manifest even at the highest energies we can create.

Creating quantum entanglement is just the first challenge— keeping it alive is where things get tricky. Imagine trying to preserve a soap bubble in a hurricane. That's what scientists face when working with quantum states. These delicate quantum connections are so fragile that even a stray air molecule or a whisper of heat can shatter them instantly. Scientists call this breakdown "decoherence," but you might think it is reality crashing in on quantum magic.

To protect these fragile quantum states, scientists have become masters of isolation. They build ultra-high vacuum chambers that make outer space look crowded, creating environments where particles can maintain their quantum dance undis-turbed. They cool their experiments to temperatures so low they make the depths of space seem tropical—just a fraction of a degree above absolute zero. The atomic jiggling that could disrupt quantum states nearly stops at these temperatures.

When even that isn't enough, they employ sophisticated error correction techniques, like musical conductors keeping an orchestra in perfect harmony despite the occasional wrong note. Some researchers work with exotic materials that naturally shelter quantum states from disruption, like building a fortress around these quantum connections. It's a constant battle against the tendency of the quantum world to lose its ethereal properties and collapse into the familiar classical behavior we see around us.

The Quantum Revolution in Security

Quantum entanglement isn't just a laboratory curiosity—it's becoming the foundation of a new type of unbreakable security. Traditional encryption relies on mathematical problems that are hard for computers to solve, but quantum encryption relies on the fundamental laws of physics themselves.

Banks have taken notice. In 2024, HSBC announced successful trials of quantum-safe technology for trading gold. JPMorgan Chase partnered with IBM to experiment with quantum algorithms for secure key exchange. These aren't just proof-of-concept experiments—they're the first steps toward a quantum-secure financial system.

China has taken the boldest step into this quantum future. In 2016, they launched Micius, the world's first quantum communication satellite. Named after an ancient Chinese philosopher, this satellite creates pairs of entangled photons in space and beams them to ground stations thousands of kilometers apart.

The achievement was remarkable. Think about what's happening: photons, created as quantum twins in space, rain down to Earth while maintaining their delicate quantum connection. 2017, this system enabled the first intercontinental quantum-encrypted video call between Beijing and Vienna. Any attempt to intercept the communication would have broken the quantum entanglement, immediately alerting both parties.

But this is just the beginning. China plans to launch a constellation of quantum satellites by 2025, creating a global network for unhackable communication. Other countries are racing to catch up. The European Union, the United States, and Japan have all announced major quantum communication initiatives.

Building the Quantum Internet

Imagine an internet where privacy is guaranteed not by software or mathematics but by the laws of physics itself. This is the promise of the quantum internet—a network that uses quantum entanglement to create unbreakable connections between users.

Unlike the classical internet, where information travels as bits (1s and 0s), the quantum internet will use quantum bits or "qubits." These qubits can be entangled with each other, creating fundamentally secure connections. Any attempt to eavesdrop would collapse the quantum state, making it impossible to intercept information without being detected.

The first quantum networks are already being built. In the

Netherlands, researchers have connected four cities with a quantum network. China has built a 2,000-kilometer quantum communication link between Beijing and Shanghai. These are the first strands of what could become a global quantum web.

But building a quantum internet can be challenging. Quantum states are fragile and difficult to maintain over long distances. Scientists are developing "quantum repeaters"—devices that can extend quantum connections without breaking them. Think of them as relay stations for quantum information, preserving the delicate quantum dance of entangled particles.

Quantum Teleportation: Science Fiction Becomes Reality

When scientists first proposed quantum teleportation in 1993, it sounded like something from Star Trek. But today, it's a laboratory reality. Using quantum entanglement, scientists can instantly transfer the quantum state of one particle to another, regardless of the distance between them.

This isn't teleportation like in science fiction—you can't beam people or objects across space. What gets "teleported" is quantum information. But this ability to instantly transfer quantum states has profound implications for quantum computing and communication.

In 2020, Fermilab scientists achieved sustained, high-fidelity quantum teleportation over 44 kilometers of fiber optics. They successfully teleported qubits with an accuracy greater than 90%. Think about what this means: quantum information disappears from one location. It reappears in another faster than any classical signal could travel between the two points.

The Philosophical Implications

As we master quantum entanglement, we're forced to confront profound questions about the nature of reality. If particles can maintain instantaneous connections across any distance, what does this mean for our concepts of space and time? If measurement at one location can instantly affect a particle elsewhere, what happens to our ideas of cause and effect?

Some physicists, like Anton Zeilinger (who won the 2022 Nobel Prize for his work on entanglement), suggest that we must fundamentally revise our understanding of reality. The universe isn't made of separate objects that interact but of relationships and connections that define what we call reality.

The Observer Effect

We've seen how particles can stay connected across vast distances, maintaining their quantum dance of possibility until we measure them. But this leads us to an even stranger aspect of quantum physics: observation itself seems to change what we observe.

The next chapter will explore how looking at quantum systems changes their behavior. We'll see how consciousness and reality might be more deeply connected than we ever imagined and how this mysterious interaction shapes everything from laboratory experiments to the possibility of quantum computers.

Remember, every time you use a secure website or tap your phone to pay, you benefit from our understanding of quantum physics. The strange quantum dance of entangled particles isn't just a scientific curiosity—it's becoming the foundation of a technological revolution that will shape our future.

Transformation: Reality's Mirror

"The moon is demonstrably not there when nobody looks." That's what physicist David Mermin said about the strange world of quantum physics. It sounds absurd—surely the moon doesn't pop in and out of existence depending on who's watching—but this bizarre idea isn't just possible at the quantum level. It's fundamental to how reality works.

When Looking Changes Reality

Imagine you're watching a street performer. They're doing a fantastic dance routine, moving with incredible freedom and creativity. But when you try to record them with your phone, they freeze into a simple, static pose. This is similar to how quantum particles behave—observing them fundamentally changes their actions.

But quantum physics takes this idea to an extreme, making our street performer analogy seem tame. In the quantum world, things don't just behave differently when observed—they exist differently. It's not just that we're disturbing delicate quantum states with clumsy measurements. Reality itself seems to transform under our gaze.

Remember our discussion of Schrödinger's cat? That thought experiment wasn't just about cats in boxes—it was highlighting something profound about the nature of reality. Before we open the box (or make any quantum measurement), the system exists in all possible states simultaneously. It's not that the cat is either alive or dead. We don't know which—according to quantum mechanics, it's actually in a superposition of both states, existing in a blur of possibilities until the moment we look.

The Measurement Problem

This brings us to physics's most profound mysteries: the measurement problem. Why does the act of observation force quantum systems to "choose" one reality from all possible options? What counts as an observation? Does it require a conscious observer, or can any interaction with the environment cause this "collapse" of possibilities into certainty?

Imagine you're dealing cards from a deck. Every day, each card is a heart, diamond, club, or spade, even before you turn it face up. But if cards behaved like quantum particles, they would exist in a mixture of all possibilities until you looked at them. Even stranger, if you could peek at just the color of the card without seeing its complete identity, it would still maintain a superposition of all possibilities consistent with that partial information.

This isn't just philosophical musing—it has real, measurable consequences. Scientists can create particles in superposition states and watch what happens when they measure them. The results consistently show that reality behaves this way whether we like it or not. As physicist Richard Feynman famously said, "Nature isn't classical, dammit, and if you want to make a simulation of nature, you'd better make it quantum mechanical."

Making Sense of the Impossible

How do we make sense of this measurement mystery? Over the decades, physicists have proposed several interpretations with mind-bending implications.

The Copenhagen Interpretation, championed by Niels Bohr and his colleagues, says we should accept that quantum systems exist in multiple states until measured. At this point, they "collapse" into one definite state. It's like saying the rules of the quantum game are fundamentally different from our everyday experience, and we shouldn't try to force our classical intuitions onto the quantum world.

However, not everyone was satisfied with this "shut up and calculate" approach. Hugh Everett III proposed something even stranger: the Many-Worlds Interpretation. In this view, when we measure, it splits into multiple branches, each containing one of the possible outcomes. Every quantum choice creates new universes, endlessly branching like an infinite cosmic tree.

If this sounds like science fiction, you're not alone. The

creators of "Avengers: Endgame" drew inspiration from these ideas. When the Avengers travel through time to collect the Infinity Stones, they're not changing their past—they're creating new branches of reality. Bruce Banner's insistence that "changing the past doesn't change the future" aligns perfectly with how quantum physicists think about parallel universes.

There's also the Pilot-Wave Theory, which suggests that particles always have definite positions and trajectories, guided by an invisible quantum wave that permeates all of space. This interpretation maintains determinism—the idea that everything has a definite cause and effect. Still, we can never directly observe without introducing a hidden layer of reality.

The Quantum Eraser: When the Present Changes the Past

When you think quantum mechanics couldn't get any stranger, enter the quantum eraser experiment. Imagine taking a photograph, but whether the picture shows a clear or blurry image depends on what you decide to do after you've already taken it. Sounds impossible? Welcome to the quantum world.

The experiment starts with our old friend, the double-slit setup. Scientists send particles through two slits, creating an interference pattern—unless they measure which slit each particle goes through, in which case the pattern disappears. But here's where it gets weird: they can "erase" this measurement information after the particles have already been detected.

221

What happens next defies common sense. When scientists erase the "which-path" information, the interference pattern reappears—even though the particles were detected before the erasure decision was made. It's as if the present decision reaches backward in time to affect the particles' past behavior.

Think about what this means. You could perform the experiment, store the data on a hard drive, and fly that drive to another continent. Months later, when you finally decide whether to erase the which-path information, that decision affects how the particles behaved in the original experiment. As physicist John Wheeler put it, "The past has no existence except as it is recorded in the present."

Time's Arrow Bends

This quantum time-bending raises profound questions about causality—the relationship between cause and effect. In our everyday world, causes always precede effects. You can't win the lottery and then buy the ticket. You can't get a sunburn and then go to the beach. But quantum mechanics suggests that, at the microscopic level, time's arrow might not be so straight.

The "grandfather paradox" famously asks what would happen if you traveled back in time and prevented your grandparents from meeting. Similar paradoxes arise naturally in quantum mechanics. The quantum eraser experiment shows that future choices can affect past events, creating what appears to be a causal loop.

Reality's Hall of Mirrors

All these quantum peculiarities force us to confront some deeply uncomfortable questions about the nature of reality itself. If particles have definite properties once we measure them, what does this mean for the world when we're not looking at them? If observation can influence the past, what does this tell us about time? And perhaps most unsettling of all: what role does consciousness play in all of this?

Some physicists, including Eugene Wigner and John Wheeler, have suggested that consciousness might play a crucial role in collapsing quantum possibilities into definite realities. Others argue that any interaction with the environment—conscious or not—can cause this collapse. The debate touches on questions that blur the line between physics and philosophy: What is consciousness? What makes a measurement a measurement? Where exactly does the quantum world end and our classical world begin?

Think about a tree falling in a forest with no one around. The old philosophical question asks if it makes a sound. Quantum mechanics raises an even stranger possibility: does the tree occupy a definite position before someone looks at it? Does it exist in a superposition of all possible states until observed? These aren't just philosophical musings—they're questions about the fundamental nature of reality itself.

The Observer Effect in Your Life

While these ideas might seem abstract, they have real implications for modern technology. Quantum cryptography uses the observer effect to create unbreakable codes—any attempt to intercept the message necessarily changes it, alerting the intended recipients to eavesdropping. Quantum computers manipulate particles in superposition states to perform calculations impossible for classical computers.

Even some biological systems might exploit quantum effects. Scientists have found evidence that birds use quantum entanglement in their magnetic navigation systems, and photosynthesis may rely on quantum coherence to achieve its remarkable efficiency. Nature has been using quantum mechanics long before we understood it.

The Ultimate Mystery

The most profound implication of quantum measurement is what it tells us about our place in the universe. We're not just passive observers watching reality unfold—we're active participants in determining what becomes real. Every measurement we make, every observation we choose to make or not make, helps shape the reality we experience.

As some New Age interpretations suggest, this doesn't mean we create reality with our minds. Instead, it means the classical world of definite properties and transparent cause-and-effect relationships emerges from a deeper quantum reality through a process we're only beginning to understand.

Quantum Technology in Action

Now that we've explored how quantum systems react to observation—how reality itself seems to transform under our gaze—it's time to see how we're putting this knowledge to work. In the next chapter, we'll explore the quantum technologies already changing our world, from quantum computers that harness superposition to perform impossible calculations to quantum sensors that can detect single atoms.

Remember, if these ideas seem strange, you're in good company. As physicist Richard Feynman once said, "I can safely say that nobody understands quantum mechanics." That may be the point. Quantum mechanics doesn't just describe the world—it reveals that it is far stranger and more mysterious than we ever imagined.

The quantum technologies we're building today aren't just clever applications of well-understood principles—they're explorations of the deepest mysteries of reality itself. Every quantum computer, every quantum sensor, and every quantum encryption system is also a philosophical experiment, probing the boundary between the observer and the observed, between the possible and the actual, and between the quantum and classical worlds.

As we move into this quantum future, we're not just developing new technologies—we're creating new ways of thinking about reality. And perhaps that's the most transformative aspect of quantum physics: it shows us that the universe is not just stranger than we imagine but stranger than we can imagine.

VI

Quantum Technology and the Future

Innovation - Tomorrow's Technology

"Nature isn't classical, dammit, and if you want to make a simulation of nature, you'd better make it quantum mechanical." Those are the words of physicist Richard Feynman, who inspired the creation of quantum computers. Today, his vision is becoming a reality—changing the world faster than you think.

From Bits to Qubits: A New Kind of Computing

Imagine trying to read a book in a pitch-dark room with just a flashlight. You can only see one word at a time, checking each word sequentially until you find what you're looking for. This is essentially how classical computers work with bits—examining information piece by piece, each bit either a 0 or 1, like a light switch on or off.

Imagine if you could flood the entire room with light, reading every word simultaneously. This is the promise of quantum computing. Instead of bits, quantum computers use qubits—quantum bits that can exist in multiple states simultaneously, like a dimmer switch on and off simultaneously.

But this analogy, while helpful, barely scratches the surface of what makes quantum computing revolutionary. A better way to think about it might be like this: Imagine trying to find your way out of an enormous maze. A classical computer would try one path at a time, backtracking when it hits dead ends. A quantum computer, through the strange properties of superposition, effectively explores all possible paths simultaneously. It's not just faster—it's fundamentally different.

The power of qubits grows exponentially. While 100 classical bits can only be in one of 2^{100} possible combinations simultaneously, 100 qubits can simultaneously exist in all 2^{100} states. That's more states than there are atoms in the visible universe. This isn't just an improvement in computing power—it's a new paradigm that defies our classical intuitions.

The Dance of Quantum Information

Creating and controlling qubits is like choreographing an impossibly delicate dance. Scientists use various physical systems to create qubits: superconducting circuits cooled to near absolute zero, individual atoms trapped by lasers, or electrons whose quantum spin can point in multiple directions simultaneously. Each approach has its challenges and advantages, but they all share one critical feature: they harness the weird properties of quantum mechanics to process information in ways classical computers never could.

Think about shuffling a deck of cards. A classical computer would have to look at each possible arrangement one at a time. But a quantum computer could consider all possible shuffles at once. With just 52 cards, there are more possible arrangements than stars in the visible universe. A quantum computer could process all these possibilities simultaneously.

This massive parallelism comes from the quantum properties discussed in previous chapters. Superposition allows each qubit to be in multiple states at once. Entanglement lets qubits share information in ways that have no classical equivalent. Even quantum tunneling, which we explored earlier, plays a role in some quantum computing architectures.

The Challenge of Quantum Coherence

But there's a catch—and it's a big one. Quantum states are incredibly fragile. The slightest interaction with the environment can cause "decoherence," destroying the delicate quantum properties that give qubits their power. It's like trying to balance a pencil on its tip while the Earth trembles from millions of tiny earthquakes.

To maintain quantum coherence, scientists go to extraordinary lengths. They cool their quantum computers to temperatures colder than deep space, just fractions of a degree above absolute zero. They use sophisticated error correction techniques to protect quantum information. They develop new materials and methods to shield their qubits from noise and interference.

Despite these challenges, quantum computers are already achieving things that would be impossible with classical computers. They're simulating complex molecular interactions for drug discovery, optimizing financial portfolios with unprecedented sophistication, and even helping us understand the quantum nature of reality itself.

Solving the Impossible: What Quantum Computers Do Best

In 2019, Google announced something extraordinary. Their quantum computer had solved a problem in 200 seconds that would have taken the world's most powerful classical supercomputer 10,000 years to crack. They called it "quantum supremacy," the moment when a quantum computer accomplishes something effectively impossible for classical computers.

But what kinds of problems are quantum computers uniquely suited to solve? They excel at tasks that classical computers find nightmarishly tricky. Imagine trying to factor a number like 2,048 into its prime components. Is it simple? Now try it with a number that's hundreds of digits long. Classical computers struggle with this task so much that we use this difficulty to encrypt much of our digital communication. Quantum computers, however, could crack such codes in minutes.

Yet code-breaking is just the beginning. In pharmaceutical laboratories, quantum computers are revolutionizing drug discovery. Traditional computers struggle to simulate complex molecular interactions—there are too many quantum effects to account for. But quantum computers speak the language of quantum mechanics natively. They can model how drugs interact with target molecules, potentially reducing the time and cost of developing new medicines from decades to years.

Nature's Quantum Calculator

Climate scientists are particularly excited about quantum computing's potential. Our current climate models, running on classical supercomputers, can only approximate molecular systems' incredibly complex quantum interactions. Quantum computers could simulate these systems, helping us understand and combat climate change more effectively.

They're also transforming material science. Want to design a better battery for electric cars, a more efficient solar panel, or a superconductor that works at room temperature? These challenges require understanding how electrons interact in complex materials—a problem tailor-made for quantum computers.

The AI Revolution Gets Quantum

Most intriguingly, quantum computers could supercharge artificial intelligence. Classical AI already seems impressive, but the sequential nature of traditional computing limits it. Quantum computers could process vast neural networks simultaneously, potentially leading to AI systems with capabilities we can barely imagine.

Think about how your phone's AI assistant recognizes your voice. It compares your speech patterns against countless possibilities until it finds a match. A quantum computer could theoretically check all possibilities simultaneously, making voice recognition instantaneous and perfect. Similar principles could revolutionize everything from image recognition to language translation.

The Quantum Internet: A New Kind of Network

While we're building ever-more-powerful quantum computers, scientists are also working on something equally revolutionary: the quantum internet. Unlike our current internet, which sends information as classical bits, the quantum internet will transmit quantum states, using entangled particles to create unhackable communication channels.

Imagine sending a message that's not just encrypted but protected by the laws of physics themselves. Any attempt to intercept the message would collapse its quantum state, immediately alerting both sender and receiver to the intrusion. China has already launched a quantum communication satellite demonstrating this technology, and other countries are racing to build their quantum networks.

But the quantum internet isn't just about secure communication. It could also connect quantum computers, creating a network of unprecedented computational power. Think of it as a quantum cloud, where the strange properties of quantum mechanics are available on demand.

Quantum Space: The Final Frontier

Space exploration is another frontier where quantum technology promises breakthroughs. Quantum sensors could detect gravitational waves with unprecedented precision, helping us understand the cosmos in new ways. Quantum navigation systems could guide spacecraft without relying on signals from Earth. Some scientists even speculate about quantum propulsion systems that could make interstellar travel possible.

NASA and other space agencies already use quantum technology to improve their measurements and communications. The exact quantum properties that make secure communication possible on Earth could one day help us stay in touch with colonies on Mars or probes in deep space.

The Near Future: What's Coming Next

We're standing at the threshold of a quantum revolution. Within the next decade, we'll likely see:

- Quantum computers solve problems in chemistry and materials science that are impossible today
- Quantum sensors detecting diseases at the molecular level before symptoms appear
- Quantum communication networks spanning continents
- Quantum financial systems that can spot market patterns invisible to classical computers

But the most exciting applications are the ones we haven't even imagined yet. When classical computers were first invented, no one predicted smartphones or social media. The quantum revolution may bring equally unexpected innovations.

A Note of Caution

Of course, with great power comes great responsibility. Quantum computers could eventually crack many encryption systems that protect our digital lives. That's why researchers are already developing "quantum-safe" encryption methods that even quantum computers can't break.

There are also concerns about quantum computing widening the technological divide between nations. Countries that master quantum technology first could gain significant advantages in everything from medicine to military applications.

The Quantum Body

While we've focused on quantum technology in this chapter, something equally fascinating awaits us: quantum biology. Your body might use quantum effects in ways we're only beginning to understand. From the efficiency of photosynthesis to the mystery of bird navigation, from the mechanics of smell to the possibility of quantum effects in consciousness, nature may have been doing quantum computing long before we invented it.

In the next chapter, we'll explore how quantum mechanics isn't just about technology—it's about life itself. We'll discover how plants use quantum coherence to turn sunlight into energy with near-perfect efficiency, how birds might use quantum entanglement to navigate using Earth's magnetic field, and how the very mechanisms of evolution might depend on quantum effects.

Remember, when Feynman said nature isn't classical, he wasn't just talking about laboratory physics experiments. He was talking about everything—including us. The quantum revolution is changing our technology and our understanding of life itself.

Medicine - Nature's Secret

"**N**ature uses only the longest threads to weave her patterns so that each small piece of her fabric reveals the organization of the entire tapestry." Richard Feynman wasn't just being poetic when he said these words—he was hinting at something profound about how the universe works, from the most minor quantum effects to the most prominent living systems.

We tend to think of quantum physics as something that only happens in laboratories, with sophisticated equipment cooled to near absolute zero. But nature has been doing quantum physics for billions of years. It's doing it right now, in every cell of your body, every leaf of every plant, and every bird's brain soaring overhead. While our most advanced quantum computers require temperatures colder than deep space to maintain quantum states for mere seconds, living things manipulate quantum systems at room temperature with remarkable precision.

The Quantum Compass in Birds' Eyes

Millions of birds embark on one of nature's most remarkable journeys every autumn. Take the Arctic tern, for instance. This small seabird navigates yearly from the Arctic to the Antarctic—a round trip of about 44,000 miles. It's like flying around the Earth's equator one and a half times, often through utterly unfamiliar territory, yet these birds rarely get lost. Even more impressive, they can navigate on cloudy days when they can't see the sun or stars and in conditions where familiar landmarks are nowhere to be seen.

For decades, this incredible feat of navigation puzzled scientists. They knew birds could sense Earth's magnetic field—a discovery that seemed like magic when first proposed. But the mechanism remained mysterious. The Earth's magnetic field is incredibly weak—about a hundred times weaker than a refrigerator magnet. How could birds detect something so subtle? And how could they use this information to determine not just direction but their position on the globe?

The answer lies in one of quantum mechanics' strangest properties: entanglement. Deep in the birds' eyes sits a protein called cryptochrome. When light hits this protein, it triggers a quantum reaction that creates a pair of entangled electrons. These electrons are exquisitely sensitive to magnetic fields thanks to a quantum property called spin. The interaction between these entangled electrons and Earth's magnetic field creates distinct quantum states that the bird's brain can interpret as directional information.

This quantum compass is so sophisticated that it shames our best navigation technology. While GPS needs at least four satellites to determine your position, birds carry a quantum positioning system in their eyes that works anywhere on Earth. They're using quantum entanglement—the same property Einstein found so troubling he called it "spooky action at a distance"—every time they migrate.

The discovery of this quantum navigation system has profound implications. First, it shows that quantum effects aren't just microscopic curiosities—they can influence behavior in large organisms. Second, it demonstrates that nature has solved one of our biggest challenges in quantum technology: maintaining quantum states at room temperature. While we struggle to keep quantum states coherent for microseconds in carefully controlled laboratory conditions, birds maintain and use quantum systems in their bodies' messy, warm environments.

Scientists are studying bird navigation to understand how nature achieves this quantum stability. The insights could help us design better quantum computers, develop more precise navigation systems, and create new medical imaging technologies. Some researchers speculate that understanding how birds process quantum information in their brains might give us clues about consciousness.

But the quantum mysteries of nature don't end with bird navigation. Even more remarkable quantum effects are at work in every green leaf and blade of grass around us, orchestrating one of life's most fundamental processes: photosynthesis.

The Quantum Dance of Photosynthesis

Every second, plants worldwide perform a miracle that powers almost all life on Earth: photosynthesis. When sunlight strikes a leaf, something remarkable happens. The light's energy must travel through a complex network of molecules to reach the reaction center, where it can be converted into chemical energy. According to classical physics, this should be inefficient, with much energy lost along the way. Yet plants achieve near-perfect efficiency in transferring this energy.

The secret? Quantum coherence. When light energy enters a leaf, instead of taking a single path through the molecular maze, it takes all possible paths simultaneously, just like a quantum particle passing through multiple slits. This quantum superposition allows the energy to find the most efficient route instantly. It's as if each photon of sunlight explores every possible path through the leaf at once, automatically choosing the best one.

Scientists discovered this quantum behavior by shooting laser pulses at photosynthetic bacteria and watching the energy move through their light-harvesting complexes. They saw something astonishing: wave-like patterns that could only be explained by quantum mechanics. The energy wasn't traveling like a classical particle bouncing through a pinball machine—it was moving like a quantum wave, spreading out and interfering with itself to find the optimal path.

What makes this even more remarkable is that it happens at room temperature, in a living cell's wet, chaotic environment. The quantum states in our best laboratories fall apart in a fraction of a second, yet plants maintain quantum coherence

241

long enough to capture sunlight efficiently. They've evolved what engineers would call a "quantum heat engine," operating with an efficiency that human technology can't match.

Nature's Quantum Technology

The implications of these discoveries stretch far beyond biology. Engineers are now studying photosynthesis to design better solar cells. If we could replicate nature's quantum efficiency, we could revolutionize renewable energy. Some scientists are even developing "artificial leaves" that use quantum effects to split water into hydrogen and oxygen, potentially creating clean fuel from sunlight.

But nature's quantum engineering continues beyond photosynthesis and bird navigation. Recent research suggests quantum effects work in our sense of smell. The traditional theory says smell works by molecules fitting into receptors like keys into locks. However, we must explain how we can distinguish between molecules with identical shapes but different isotopes. The answer might lie in quantum tunneling—the same effect that makes transistors work in your computer might help you distinguish the scent of a rose from a lily.

Even enzymes, the proteins that catalyze chemical reactions in our bodies, might use quantum tunneling to speed up their work. When an enzyme needs to transfer a hydrogen atom from one molecule to another, sometimes the atom doesn't go over the energy barrier—it tunnels through it, just like the quantum particles we discussed earlier.

The Future of Quantum Medicine

While plants and birds have been using quantum effects for millions of years, we're just beginning to harness quantum mechanics for human health. A new field called quantum medicine is emerging, promising to revolutionize how we diagnose and treat diseases.

Consider quantum imaging technology. Traditional MRI machines already use quantum effects—the spin of atomic nuclei—to create images of our bodies. However, new quantum sensors could detect individual molecules that are out of place, potentially spotting diseases like cancer years before they become visible through conventional methods. It's like giving doctors quantum glasses that let them see the earliest signs of illness at the molecular level.

Drug discovery is another field being transformed by quantum science. Understanding how molecules interact is fundamentally a quantum mechanical problem—electrons jumping between energy levels and atoms sharing quantum states to form chemical bonds. Classical computers struggle to simulate these interactions accurately, so developing new drugs traditionally takes years and billions of dollars. Quantum computers, however, speak the language of quantum mechanics natively. They could simulate molecular interactions with unprecedented accuracy, potentially reducing drug development time from decades to months.

Some scientists are even exploring how quantum effects might help target cancer cells more precisely. Traditional radiation

243

therapy can damage healthy tissue along with cancer cells. However, quantum-controlled radiation could theoretically target diseased cells with atomic precision, minimizing side effects and maximizing treatment effectiveness.

The Quantum Brain

The most intriguing frontier in quantum biology is the possibility that quantum effects play a role in how our brains work. The human brain contains roughly 86 billion neurons connected to thousands of others, creating a network of unimaginable complexity. Could quantum mechanics help explain how this intricate system gives rise to consciousness?

Nobel laureate Roger Penrose and anesthesiologist Stuart Hameroff have proposed that quantum processes in structures called microtubules—tiny tubes that help maintain cellular structure—might be crucial to consciousness. While their specific theory remains controversial, the idea that quantum effects might influence brain function is gaining scientific attention.

Recent research has found evidence of quantum coherence in the brain's molecular machinery. The exact quantum effects that help plants harvest sunlight efficiently might help our neurons process information. Some scientists speculate that quantum entanglement between molecules in different neurons could help explain the brain's remarkable information-processing abilities.

Nature's Quantum Legacy

244

As we've seen throughout this chapter, nature didn't just stumble upon quantum mechanics—it mastered it. From the quantum compass in birds' eyes to the quantum coherence in photosynthesis, from the possible quantum effects in our sense of smell to the mysteries of consciousness, living things have been exploiting quantum phenomena for billions of years.

This realization is transforming how we think about biology and medicine. We're learning that life isn't just subject to quantum effects—it actively harnesses them. Each discovery of quantum processes in nature provides new insights that could help us develop better technologies and treatments.

The Quantum Universe

We've seen how quantum physics shapes life on Earth, from the tiniest cellular processes to the possibility of quantum effects in consciousness. But the quantum story is much more significant. In the next chapter, we'll see how quantum mechanics shapes life and the entire universe—from the nuclear furnaces of stars to the vast cosmic web of galaxies.

Remember, every breath you take, every beat of your heart, every thought in your mind might depend on quantum effects that we're only beginning to understand. The quantum world isn't just about subatomic particles in physics labs—it's about us, life itself, and our place in this remarkably quantum universe.

Expansion - The Quantum Universe

"What we have called matter is energy, whose vibration has been so lowered as to be perceptible to the senses. There is no matter." When Einstein made this statement, he was hinting at something profound: the universe, at its deepest level, isn't made of tiny billiard balls we call particles. Instead, reality emerges from something far stranger—a cosmic quantum dance of fields and energy that stretches across space and time.

The Symphony of Fields

Imagine sitting by a still pond on a quiet morning. Drop a pebble, and waves ripple across the water's surface, creating circular patterns that spread outward. Drop two pebbles, and their ripples interfere, creating complex patterns of peaks and troughs. Now imagine the entire universe filled not with water but invisible fields permeating every point in space. These aren't like magnetic or electric fields you might remember from school—they are quantum fields and the foundation of everything we see around us.

According to quantum field theory, our universe is more like a vast collection of overlapping symphonies than a container full of particles. Just as a symphony hall fills with overlapping sound waves from different instruments, space fills with overlapping quantum fields. Each type of particle we observe—electrons, photons, quarks—corresponds to its universal field that extends throughout space and time.

Particles are localized vibrations in these fields, like notes played on cosmic instruments. When we observe an electron, we're not seeing a tiny sphere of matter—we're detecting a particular kind of vibration in the electron field, much like hearing a specific note in a symphony. When we see light, we're not observing little particles streaming through space—we're detecting oscillations in the electromagnetic field.

These quantum fields interact and influence each other in precise mathematical ways. When an electron seems to emit or absorb a photon, the electron field and electromagnetic field exchange energy, creating what we perceive as particle interactions. It's like two musical instruments playing together, creating harmonies and resonances that we interpret as physical events.

The mathematics of quantum field theory makes precise predictions about how these fields interact. It successfully explains phenomena that baffled earlier physicists, like how particles can pop into existence from apparent space or how forces can act across distances. The theory is so accurate that some of its predictions have been confirmed to a precision of one part in a trillion—like measuring the distance between New York and Los Angeles to the width of a human hair.

This view of reality revolutionizes our understanding of matter itself. The solid objects we see around us—tables, chairs, even our bodies—are not fundamentally different from the forces that act between them. Both matter and forces emerge from the same underlying quantum fields. The distinction between substance and energy, particle and wave, matter and force dissolves at this most profound level of reality.

The discovery of the Higgs boson in 2012 dramatically confirmed quantum field theory. The Higgs field was predicted to exist because the mathematics of quantum fields required it—without it, other particles couldn't have mass. When the Large Hadron Collider detected the predicted vibration in this field (the Higgs boson), it validated our understanding of how quantum fields create the properties of matter we observe.

The Quantum Nature of Nothing

What's in space? Our common sense tells us "nothing," but quantum field theory reveals something more fascinating. Even in the emptiest region of space, quantum fields are never truly still. They fluctuate constantly, like a calm ocean that's never wholly flat but always has tiny ripples moving across its surface.

These quantum fluctuations have remarkable consequences. According to the uncertainty principle we discussed earlier, we can never know the strength of a field and how quickly it changes simultaneously. This means that even in a vacuum, fields must maintain minimal activity—a sort of quantum jitter that can never be eliminated.

This isn't just theoretical speculation. These vacuum fluctuations have measurable effects in the real world. Consider the Casimir effect, first predicted by Dutch physicist Hendrik Casimir in 1948. Place two metal plates extremely close together in a vacuum, and they'll experience a tiny attractive force. Why? The quantum fluctuations between the plates are restricted to specific wavelengths, while all wavelengths are possible outside the plates. This difference in quantum activity creates a small but measurable pressure pushing the plates together.

Even more remarkably, these vacuum fluctuations can momentarily create pairs of particles and antiparticles that pop into existence and then quickly annihilate each other. Physicists call these "virtual particles," while they exist too briefly to be directly observed, their effects can be measured. They contribute to the properties of atoms, affect the strength of electromagnetic forces, and might even play a role in the behavior of black holes through Hawking radiation.

Think about what this means: the vacuum isn't empty at all. It's a seething quantum foam of virtual particles appearing and disappearing, fields fluctuating between positive and negative values, and energy borrowing itself from nothing for the briefest moments allowed by the uncertainty principle. Every cubic centimeter of "empty" space contains more activity than we could imagine.

Dark Mysteries and Quantum Solutions

THE BEGINNER'S GUIDE TO REALITY

This quantum view of space might help us understand two of the biggest mysteries in modern cosmology: dark matter and dark energy. These invisible components make up 95% of the universe's content, yet we still don't know what they are.

Dark matter reveals itself through its gravitational effects on galaxies and galaxy clusters. It forms an invisible scaffold that holds cosmic structures together. While we can't see it directly, some theorists suggest it might be made of particles that interact through yet-undiscovered quantum fields. These hypothetical particles rarely interact with ordinary matter except through gravity, making them extremely difficult to detect.

Dark energy is even more mysterious. It is a property of space itself, pushing the universe to expand at an accelerating rate. Some physicists speculate that dark energy might be related to vacuum energy—the energy of quantum fluctuations in space. The challenge is that when we calculate how much vacuum energy there should be using quantum field theory, we get a number that's way too large. This "cosmological constant problem" remains one of physics's most giant unsolved puzzles.

When Reality Multiplies

The quantum nature of the universe raises an even more mind-bending possibility: our universe might be just one of many. Remember the Many-Worlds Interpretation of Quantum Mechanics we discussed earlier? It suggests that every quantum event creates new branches of reality. But quantum field theory takes this idea even further, hinting at the possibility of multiple universes emerging from the quantum vacuum itself.

Think about those virtual particles we discussed, popping in and out of existence in space. Some theories suggest that similar quantum fluctuations, operating at a cosmic scale, might give birth to entire universes. Each would bubble up from the quantum vacuum like a bubble forming in boiling water, expanding to create its own space, time, and physical laws.

String theory, an ambitious attempt to unite quantum mechanics with gravity, makes this multiverse idea even more intriguing. It suggests that we think of particles as tiny vibrating strings of energy, like infinitesimal violin strings playing cosmic melodies. These strings would vibrate in extra dimensions we can't see, creating all the particles and forces we observe.

Just as a guitar string can vibrate in different patterns to produce different notes, these cosmic strings could vibrate in countless ways, potentially creating various particles and forces in other universes. With its particular set of physical laws and constants, our universe might be just one possible "song" these strings can play.

If you've watched "Doctor Strange in the Multiverse of Madness," you've seen a Hollywood version of these ideas. While the movie takes creative liberties, the core concept—that reality might be far more expansive than our single universe—

emerges naturally from our best physical theories.

The Quantum Tapestry

A remarkable picture emerges as we zoom out from quantum fields to multiple universes. The exact quantum principles we discovered by studying tiny particles shape the most significant structures in the cosmos. The uncertainty principle affecting electron positions helps create new universes, and the vacuum fluctuations influencing atomic properties drive cosmic expansion.

Every atom in your body exists as a set of vibrations in quantum fields that extend throughout space and time. These fields interact according to precise mathematical rules, creating the intricate dance we call reality. The space around you teems with virtual particles and quantum fluctuations. And our entire universe might be one bubble in an infinite quantum foam of possible realities.

This view of the cosmos raises profound questions. If our universe is one of many, what does this mean for the laws of physics? Are they universal constants or just local bylaws? What happens to all those other possibilities if reality branches every time a quantum choice is made? And perhaps most fundamentally, why does our universe have the properties that allow for our existence?

The Nature of Reality

We've traveled from the tiniest quantum particles to the

most significant structures in the universe. We've seen how quantum physics shapes everything—from the atoms in your body to the stars in the sky, from the vacuum of space to the possibility of multiple universes. Now, it's time to step back and ask what this all means for our understanding of reality.

In the final chapter, we'll explore the philosophical implications of quantum physics. We'll confront questions about the nature of reality, the role of consciousness in the quantum world, and what quantum mechanics tells us about our place in this vast, strange, and beautiful cosmos.

Remember, when Einstein said there is no matter, he wasn't being metaphorical. The solid reality we experience emerges from an underlying quantum realm that challenges our deepest intuitions about what is real. As we've seen throughout this book, the universe is not only stranger than we imagine—it's stranger than we can imagine. But through quantum physics, we're beginning to glimpse the true nature of this cosmic quantum dance.

Conclusion: Your Quantum Journey

When we began this journey through quantum physics, we started with a simple idea: that the universe, at its deepest level, behaves in ways our everyday experience could never prepare us for. Having explored everything from the double-slit experiment to quantum computers, from entangled particles to the possibility of multiple universes, we can see just how profound this insight truly is.

Quantum physics isn't just about tiny particles or abstract mathematics—it's about the fundamental nature of reality itself. Every photon in a sunbeam, an electron in a lightning bolt, and an atom in your body follows quantum rules. The solid world we experience emerges from a deeper quantum reality where particles can exist in multiple places simultaneously, where space teems with virtual particles, and where the mere act of observation can change the outcome of an experiment.

But most remarkably, we've seen how nature harnesses quantum effects with extraordinary precision. Birds navigate using quantum entanglement in their eyes. Plants achieve near-perfect efficiency in photosynthesis through quantum coherence. Even our consciousness might depend on quantum processes in our brains. While we struggle to maintain quantum states for microseconds in our most advanced laboratories, living things have been doing quantum physics for billions of years.

The technologies we build based on these quantum principles are already transforming our world. Quantum computers are tackling problems that classical computers could never solve. Quantum cryptography promises unbreakable codes protected by the laws of physics themselves. Quantum sensors might soon detect diseases at the molecular level, revolutionizing medicine.

We've broken down these complex ideas into understandable pieces through the S.P.A.C.E.T.I.M.E. framework. You now grasp concepts that once seemed impossible to comprehend. You understand why Einstein called entanglement "spooky action at a distance," Schrödinger put his famous cat in a box, and Feynman said, "Nobody understands quantum mechanics."

But this isn't the end of your quantum journey—it's just the beginning. Scientists are making discoveries about how quantum mechanics shapes our world daily. As you go forward, keep asking questions. Stay curious about the quantum nature of reality. Remember that every technology you use, from your smartphone to your GPS, works because someone dared to explore the quantum realm.

Please leave a review if you've found this book helpful in

understanding the quantum world. Your feedback helps other curious minds navigate these fascinating ideas.

The universe is far stranger and more wonderful than we ever imagined. But we're beginning to understand its most profound mysteries through quantum physics. Keep exploring, keep learning, and keep marveling at the quantum dance that underlies all of reality.

Glossary: Understanding the Language of Quantum Physics

Amplitude: In quantum mechanics, a complex number that determines the probability of finding a particle in a particular state. The square of the amplitude's magnitude gives the likelihood of measuring that state.

Antiparticle: A particle with the same mass but opposite properties (like charge) as its regular counterpart. When particles meet their antiparticles, they annihilate each other, converting their mass into pure energy.

Bell's Theorem: A mathematical proof by John Bell showing that quantum mechanics predicts stronger correlations between particles than any theory based on local hidden variables could explain. It demonstrates that quantum entanglement is genuine and can't be explained by classical physics.

Bohr Model: An early model of the atom proposed by Niels Bohr, suggesting that electrons orbit the nucleus in fixed energy levels, like planets around the sun. While superseded by quantum mechanics, it helped establish quantized energy states.

Bose-Einstein Condensate: A state of matter formed when particles called bosons are cooled to temperatures close to absolute zero. In this state, multiple particles occupy the same quantum state, behaving as a single quantum entity.

Copenhagen Interpretation: This is the traditional interpretation of quantum mechanics, developed primarily by Niels Bohr and Werner Heisenberg. It states that quantum systems exist in all possible states simultaneously until measured. At this point, they "collapse" into a definite state.

Coherence: The property that allows quantum systems to exist in multiple states simultaneously. When a system is coherent, its quantum waves remain in step and can interfere with each other, like ripples on a pond.

Dark Energy: A mysterious form of energy that appears to permeate all of space and drives the universe's accelerating expansion. It might be related to quantum vacuum energy.

Dark Matter: An invisible form of matter that interacts through gravity but not through electromagnetic forces. It might be composed of yet-undiscovered quantum particles.

Decoherence: The process by which quantum systems lose their quantum properties through interaction with their environment. It explains why we don't see quantum effects in everyday objects.

Eigenstate: A quantum state with a definite value for a particular observable property. When measured, a system in an eigenstate will always yield the same result for that property.

Entanglement: A quantum connection between particles where measuring one instantly determines the state of the other, no matter how far apart they are. Einstein called this "spooky action at a distance."

Fermion: A type of particle following the Pauli exclusion principle, meaning no two fermions can occupy the same quantum state. Electrons, protons, and neutrons are all fermions.

Field: In quantum field theory, a field exists at every point in space and can oscillate like a wave. Particles are excitations in these fields, like waves on an ocean.

Gauge Theory: A type of quantum field theory that describes force-carrying particles (like photons) arising from natural symmetries.

Heisenberg Uncertainty Principle: A fundamental limit of nature stating that we cannot simultaneously know both a particle's position and momentum with perfect precision. The more accurately we measure one, the less we know about the other.

Interference: The phenomenon where quantum waves combine, reinforcing or canceling each other. This creates patterns like those seen in the double-slit experiment, demonstrating the wave nature of quantum objects.

THE BEGINNER'S GUIDE TO REALITY

Many-Worlds Interpretation: A theory proposing that quantum measurements don't collapse the wave function but instead split reality into multiple branches, each containing one possible outcome. According to this interpretation, all possible quantum outcomes occur in different parallel universes.

Measurement Problem: The fundamental question in quantum mechanics is why and how quantum systems appear to change their behavior when measured. This problem lies at the heart of debates about quantum interpretation.

Observer Effect: The principle that measuring a quantum system inevitably affects it. This is different from the uncertainty principle—it's about the physical interaction required to make a measurement.

Orbital: In quantum mechanics, a region of space where an electron is likely to be found around an atom. Unlike the circular orbits of the Bohr model, orbitals are probability clouds with complex three-dimensional shapes.

Pauli Exclusion Principle: A fundamental rule stating that no identical fermions (like electrons) can simultaneously occupy the same quantum state. This principle explains the atomic structure and why the matter is solid.

Photon: A quantum of electromagnetic radiation—the particle of light. Photons have no mass, always travel at the speed of light, and demonstrate wave and particle properties.

Pilot Wave Theory: An alternative interpretation of quantum mechanics proposing that particles have definite positions and are guided by an actual wave. This theory maintains determinism but requires instantaneous action at a distance.

Planck Constant: A fundamental physical constant (h) that defines the scale at which quantum effects become important. It relates a particle's energy to its wavelength and is crucial in many quantum mechanical equations.

Planck Length: The minor meaningful length scale in physics (about ten^{-35} meters), where quantum effects and gravity become equally important. Below this scale, our current physics theories break down.

Probability Amplitude: A complex number used in quantum mechanics whose square gives the probability of measuring a particular outcome. Unlike classical probabilities, amplitudes can be harmful or complex numbers.

Probability Cloud: A region of space where a quantum particle is likely to be found when measured. The denser the cloud at a particular point, the more likely the particle will be detected.

Quantum Chromodynamics (QCD): The theory describing the strong nuclear force that binds quarks together to form protons, neutrons, and other particles. It explains why quarks are never found alone but always in combinations.

Quantum Cryptography: A secure communication method

that uses quantum mechanics principles, particularly entanglement and the observer effect, to detect eavesdropping attempts. Unlike classical encryption, its security is guaranteed by the laws of physics.

Quantum Electrodynamics (QED): The quantum theory of the electromagnetic force, describing how light interacts with matter. It's one of science's most precisely tested theories, accurately predicting electromagnetic effects to within one part in a trillion.

Quantum Entanglement: When two or more particles become correlated so that the quantum state of each cannot be described independently. Measuring one particle instantly determines the state of its entangled partners, regardless of distance.

Quantum Field: A field that permeates all of space and can exhibit quantum mechanical properties. Different types of particles are understood as excitations in their corresponding quantum fields, like ripples on the surface of an ocean.

Quantum Fluctuation: A temporary change in the amount of energy at a point in space, as the uncertainty principle allows. These fluctuations explain the Casimir effect and may have triggered the Big Bang.

Quantum Number: A value that describes a specific property of a quantum system, such as energy level, angular momentum, or spin. Quantum numbers determine what states are possible for particles in atoms.

Quantum Superposition: The ability of a quantum system to exist in multiple states simultaneously until measured. This principle underlies phenomena like quantum computation and quantum tunneling.

Quantum Teleportation: A technique using quantum entanglement to transmit quantum information from one location to another. Despite its name, it doesn't transport matter or energy, only quantum states.

Quantum Tunneling: A phenomenon where particles pass through energy barriers that they classically couldn't overcome. This effect explains radioactive decay and makes specific electronic devices possible.

Qubit: The quantum equivalent of a classical computer bit. Unlike classical bits that are either 0 or 1, qubits can exist simultaneously in a superposition of both states, enabling quantum computation.

Schrödinger Equation: The fundamental equation of quantum mechanics that describes how quantum states evolve. It plays a role similar to Newton's laws in classical mechanics.

Spin: A quantum property of particles that has no classical analog. Despite its name, it's not about physical rotation. Still, it is instead an intrinsic form of angular momentum that can only take specific discrete values.

String Theory: A theoretical framework attempting to unify quantum mechanics with gravity by proposing that all particles and forces are different vibrations of tiny one-dimensional "strings."

Vacuum Energy: The energy present in space due to quantum fluctuations. This energy might be related to dark energy and the universe's expansion.

Wave Function: The mathematical description of a quantum system that contains all possible information about its state. The square of its amplitude gives the probability of measuring particular properties.

Wave-Particle Duality: The principle that all matter and energy exhibit wave-like and particle-like properties, demonstrating fundamentally different behavior depending on how they're observed or measured.

This completes our comprehensive glossary of quantum physics terms. These concepts form the foundation for understanding the strange quantum world and its applications in modern technology. While some terms might seem abstract, remember that they describe natural phenomena that scientists observe and use daily in laboratories.

Your Review: A Glitch in the Matrix - Shape All Three Realities

Has this trilogy changed how you think about existence? Whether quantum physics made you question the nature of particles, parallel universes expanded your view of possibility, or simulation theory left you wondering about the fabric of our reality - your journey through all three books could light the way for others.

Take a moment to leave a review wherever you purchased the books. Whether you're now explaining quantum entanglement at dinner parties, debating the many-worlds interpretation, or considering if we live in a sophisticated computer simulation, your experience matters. Your words could be the beacon that guides another curious mind through these three fascinating knowledge domains.

Each book in this series - from quantum physics to parallel universes to simulation theory - builds upon the last to create a complete picture of our possible reality. By sharing your thoughts on the complete trilogy, you'll help others decide to embark on the same mind-expanding journey.

Consider your review creating a new branch in the multiverse

- one where another reader discovers these fascinating ideas thanks to your shared experience. Writing a review takes just a few minutes, but its impact could ripple across reality itself.

Thank you for being part of our community of reality explorers. Your perspective helps us continue making these complex concepts accessible to everyone curious about the true nature of our existence.

www.ingramcontent.com/pod-product-compliance
Lightning Source LLC
Chambersburg PA
CBHW071549210326
41597CB00019B/3173